Performance and Metal Release of Non-Leaded Brass Meters, Components, and Fittings

Awwa
Research
Foundation

About the Awwa Research Foundation

The Awwa Research Foundation (AwwaRF) is a member-supported, international, nonprofit organization that sponsors research to enable water utilities, public health agencies, and other professionals to provide safe and affordable drinking water to consumers.

The Foundation's mission is to advance the science of water to improve the quality of life. To achieve this mission, the Foundation sponsors studies on all aspects of drinking water, including supply and resources, treatment, monitoring and analysis, distribution, management, and health effects. Funding for research is provided primarily by subscription payments from approximately 1,000 utilities, consulting firms, and manufacturers in North America and abroad. Additional funding comes from collaborative partnerships with other national and international organizations, allowing for resources to be leveraged, expertise to be shared, and broad-based knowledge to be developed and disseminated. Government funding serves as a third source of research dollars.

From its headquarters in Denver, Colorado, the Foundation's staff directs and supports the efforts of more than 800 volunteers who serve on the board of trustees and various committees. These volunteers represent many facets of the water industry, and contribute their expertise to select and monitor research studies that benefit the entire drinking water community.

The results of research are disseminated through a number of channels, including reports, the Web site, conferences, and periodicals.

For subscribers, the Foundation serves as a cooperative program in which water suppliers unite to pool their resources. By applying Foundation research findings, these water suppliers can save substantial costs and stay on the leading edge of drinking water science and technology. Since its inception, AwwaRF has supplied the water community with more than $300 million in applied research.

More information about the Foundation and how to become a subscriber is available on the Web at **www.awwarf.org.**

Performance and Metal Release of Non-Leaded Brass Meters, Components, and Fittings

Prepared by:
Anne Sandvig, **Glen Boyd**, and **Gregory Kirmeyer**
HDR Engineering
500 108th Ave. N.E., Suite 1200, Bellevue, WA 98004-5549

Marc Edwards and **Simoni Triantafyllidou**
Virginia Polytechnic Institute and State University
418 New Engineering Bldg, Mail Code 0246, Blacksburg, VA 64061

and

Brian M. Murphy
Black and Veatch
4800 Meadows Road, Suite 200, Lake Oswego, OR 97035

Jointly sponsored by:
Awwa Research Foundation
6666 West Quincy Avenue, Denver, CO 80235-3098

and

U.S. Environmental Protection Agency
Washington D.C.

Published by:

Awwa Research Foundation

American Water Works Association

IWA Publishing

DISCLAIMER

This study was jointly funded by the Awwa Research Foundation and the United States Environmental Protection Agency (USEPA) under cooperate agreement CR-83110401. AwwaRF and USEPA assume no responsibility for the content of the research study reported in this publication or for the opinions or statements of fact expressed in the report. The mention of trade names for commercial products does not represent or imply the approval or endorsement of AwwaRF or USEPA. This report is presented solely for information purposes.

CONTENTS

TABLES

LIST OF FIGURES

FOREWORD

The Awwa Research Foundation is a nonprofit corporation that is dedicated to the implementation of a research effort to help utilities respond to regulatory requirements and traditional high-priority concerns of the industry. The research agenda is developed through a process of consultation with subscribers and drinking water professionals. Under the umbrella of a Strategic Research Plan, the Research Advisory Council prioritizes the suggested projects based upon current and future needs, applicability, and past work; the recommendations are forwarded to the Board of Trustees for final selection. The foundation also sponsors research projects through the unsolicited proposal process; the Collaborative Research, Research Applications, and Tailored Collaboration programs; and various joint research efforts with organizations such as the U.S. Environmental Protection Agency, the U.S. Bureau of Reclamation, and the Association of California Water Agencies.

This publication is a result of one of these sponsored studies, and it is hoped that its findings will be applied in communities throughout the world. The following report serves not only as a means of communicating the results of the water industry's centralized research program but also as a tool to enlist the further support of the nonmember utilities and individuals.

Projects are managed closely from their inception to the final report by the foundation's staff and large cadre of volunteers who willingly contribute their time and expertise. The foundation serves a planning and management function and awards contracts to other institutions such as water utilities, universities, and engineering firms. The funding for this research effort comes primarily from the Subscription Program, through which water utilities subscribe to the research program and make an annual payment proportionate to the volume of water they deliver and consultants and manufacturers subscribe based on their annual billings. The program offers a cost-effective and fair method for funding research in the public interest.

A broad spectrum of water supply issues are addressed by the foundation's research agenda: resources, treatment and operations, distribution and storage, water quality and analysis, toxicology, economics, and management. The ultimate purpose of the coordinated effort is to assist water suppliers in providing the highest possible quality of water economically and reliably.

Walter J. Bishop
Chair, Board of Trustees
Awwa Research Foundation

Robert C. Renner, P.E.
Executive Director
Awwa Research Foundation

ACKNOWLEDGMENTS

The authors would like to acknowledge the gracious support of the Awwa Research Foundation (AwwaRF) and the United States Environmental Protection Agency (USEPA) without whose support, this project would not have been possible.

The authors wish to extend their thanks and appreciation to Traci Case, AwwaRF Project Manager and the following AwwaRF Project Advisory Committee members for their expertise and constructive guidance and contributions throughout the project:

- David Heumann, Los Angeles Department of Water and Power
- Jeff Kempic, United States Environmental Protection Agency
- Dan Laprade, Massachusetts Department of Environmental Protection

Several utilities, agencies, and companies from across the country provided significant time, staff, and expertise, as participants in the project's Expert Workshop. The authors gratefully acknowledge their efforts:

- Rich Brown, Portland Water Bureau
- Bob Powell, Pinellas County Water
- Jack DeMarco, Greater Cincinnati Water Works
- Michael Schock USEPA, National Risk Management Research Laboratory
- Lisa Donahue, USEPA Region 3
- France Lemieux, Health Canada
- Pete Greiner, NSF International
- Clif McLellan, NSF International
- Jeremy Brown , NSF International
- Dave Viola, Plumbing Manufacturer's Institute
- Bob Weed, Copper Development Association
- Andy Kireta, Copper Development Association
- Lance Agness, Ford Meter Box
- Mike Jones, Ford Meter Box
- Craig Selover, Masco Corporation
- Lee Mercer, Moen Incorporated
- Brad Homiston, Kohler Company
- Todd Kidder, Federal Metal Company
- Joseph Napolitan, Extruded Metals Company
- Barry Munce, Extruded Metals Company
- Dom Lemaire, Concast Metal Products Co.
- Harry H. Libengood, Jr., Concast Metal Products Co.
- Richard J. Emerick, Elkay Manufacturing Company
- Devin K. Denner, Chase Brass
- Larry Muller, Chase Brass
- Myron Lee, Mueller Brass

- Mike Buyarski, Colonial Metals
- Tim Bianchi, Neptune Meter
- Don Kullman, Neptune Meter

The authors would also like to extend a special thanks to the Copper Development Association who provided a financial contribution to the project. Finally, we would like to thank Diane Meister for her skillful coordination of the Expert Workshop and Julie Self for her efforts in preparing the final report.

EXECUTIVE SUMMARY

INTRODUCTION

This project was designed to provide definition and structure to the issues surrounding the current state of knowledge, testing protocols, performance, regulatory environment, and research gaps pertaining to the widespread use of residential sized non-leaded brass materials in potable water systems. Historically, leaded brasses have been used in potable water systems, however they have been found to leach lead under certain circumstances and therefore alternative non-leaded brasses have been developed. Non-leaded brass is defined as brass with no lead added, which typically equates to a lead content of between 0.1% to 0.25% by weight. The goal of the project was to synthesize the current state of knowledge through a series of topic papers and then convene experts in the field to develop research needs and priorities.

To achieve that goal, a review and summary of the current understanding of non-leaded brass plumbing and distribution components were completed, the major issues related to their manufacture, specification, and use in potable water distribution systems were identified, and a plan was prepared to address research data gaps and practical information related to their use. In completing this project, the research team addressed the following key issues:

1. Composition of non-leaded alloys versus leaded alloys
2. Mechanical properties of non-leaded alloys compared to leaded alloys
3. Manufacturing implications of non-leaded components
4. Potential for metals leaching from non-leaded components (lead, copper, bismuth, selenium) and the health effects associated with these compounds
5. Alternatives to non-leaded materials for fabricating appurtenances
6. Expected field performance and corrosion resistance of non-leaded components
7. Potential impacts to water utility operations and maintenance activities associated with non-leaded brass components
8. Motivations for utilities to replace existing inventories with non-leaded materials
9. Potential plumbing codes, standards, inspection, and enforcement impacts of moving the plumbing industry towards non-leaded components

How was this study conducted?

The research team's approach to this project was to develop a series of topic papers that identified the issues and knowledge gaps related to the implementation of non-leaded components in the distribution system. These topic papers included:

- Description of the Composition, Lead Content, and Mechanical Properties of New Non-Leaded Brasses
- Potential Health Effects of Non-Leaded Brass Alloys
- Potential on Non-Leaded Brasses to Release Lead and Copper and Other Metals (Bismuth and Selenium) into Drinking Water
- The American Society for Testing and Materials (ASTM) Specifications for New Non-Leaded Brasses

- Summary of Current State Experience with Non-Leaded Materials
- Utility Motivations and Experience with Non-Lead Brass Components
- Potential Impacts to Manufacturers
- Impacts to Plumbing Industry
- Impacts to Utility Operations and Maintenance Procedures
- Alternative Materials for Fittings and Components.

The findings of the topic papers were then used as a framework to focus the discussion of a two-day Expert Workshop. The objectives of the workshop were to:

- Identify, synthesize, and prioritize the issues surrounding non-leaded materials as identified by the topic paper.
- Identify, prioritize, and describe research gaps and needs.
- Develop a recommended research plan to fill the data gaps in the areas of performance, testing, and impacts on drinking water and other areas as identified.

These objectives were accomplished by a panel of stakeholders representing utilities, manufacturers, regulators, academicians, professional organizations and testing/certification organizations with experience regarding the potential use of non-leaded components in potable water systems. These participants brought to the table knowledge related to drinking water regulation, manufacturing, water utility operations and maintenance, cost, performance, materials testing, and research. The findings of the expert panel are the basis for the research plan described in this report.

Why is this project important to utilities?

Water utilities throughout the U.S. have begun to specify non-leaded materials for their systems. However, there are several knowledge gaps associated with the manufacture and performance of in-line and premise plumbing components made using these new non-leaded materials. Water utilities need to understand the performance of these non-leaded components in the field, the associated operations and maintenance issues that may arise with their use, and their potential for leaching other compounds into the drinking water. There is also a need to understand the process and timeframes for adjusting standards and specifications to address these components.

What is the difference between leaded, lead-free, and non-leaded?

Leaded materials are those solders, fluxes, brasses, pipes, and fittings typically installed prior to their 1986 ban. These materials contain a varying amount of lead as a percent of their total composition. The 1986 Safe Drinking Water Act amendments banned solder and flux containing more than 0.2 percent lead and any pipe and plumbing fitting or fixture that was not "lead-free." That legislation defined l*ead-free* as a material having a lead content of 8 percent or less of total content. While the regulatory term is 'lead-free", these components can still contain up to 8 percent lead. New non-leaded brass alloys contain between 0.1 to 0.25 percent lead. No lead is added to these alloys. The minimal amount of lead present is due to contributions from recycled materials and ore that are used as the source metals. These alloys, and components manufactured using these alloys, will be referred to as "non-lead" or non-leaded" in this report.

What are the identified research priorities?

A total of 27 priority issues and research needs were identified, generally categorized into the following major areas:

Leaching, Performance, and Product Acceptance Criteria. The importance of understanding the leaching and corrosion potential of non-leaded brass components, and their consequent short- and long-term field performance in utility settings were re-current issues identified during the Expert Workshop.

Costs. Identifying the potential costs differentials associated with the use of non-leaded components, not only material costs but overall program costs, would provide valuable information to utilities, as would the costs related to raw materials supply and recycling for the manufacturing industry.

Plumbing Code Revisions and Marketplace Enforcement. The standardization and use of certified products for water distribution would be aided by determining what steps should be taken to develop consistent and effective plumbing codes, incorporating non-leaded brass components into current codes, and identifying certification and enforcement programs to ensure the use of certified products.

Education and Technology Transfer. There is a need to develop and implement educational and technology transfer programs to share information between the various stakeholders involved in the manufacture, use, and regulation of non-leaded brass components. Communicating the costs, benefits, and constraints of using non-leaded components, understanding terminology differences amongst the various groups, and providing information on the health benefits of utilizing non-leaded materials were identified as possible topics.

Manufacturing Issues. Several research needs were identified related to the manufacture of non-leaded alloys and components. Developing information on global supply and demand of the bismuth and selenium used to manufacture non-leaded alloys, the incidental impurities in recycled materials and manufacturing difficulties associated with dissimilar scrap streams, were all identified as information needed by the manufacturing community.

Regulatory Impact and Implementation. Several priority issues were identified related to incorporating non-leaded brass components into the current regulatory approach. Developing information on occurrence, toxicology, and chemical characteristics of the elements used to manufacture non-leaded brass (primarily bismuth and selenium) were identified, as was the need for risk assessments for these elements. Additional needs included developing federal regulatory approaches for non-leaded brass (such as specifying lead content in materials, requiring health based performance standards, or regulating products at the point of sale), assisting states in developing and implementing programs related to non-leaded brass, and addressing inconsistencies between federal and state regulations.

What are the applications of the project's findings?

This report provides the water supply community with:

- A summary of current knowledge on the composition, mechanical properties, and ASTM specification of non-leaded brasses, as well as a summary of the leaching and corrosion potential of non-leaded brasses, and the potential health effects of these leachates (lead, copper, selenium, bismuth, etc.). This information provides utilities with a basic understanding of non-leaded brass alloys used in the manufacture of drinking water components.
- A preliminary assessment of the performance of non-leaded brass alloys and costs associated with replacement of traditional leaded components with non-leaded components. This information is particularly useful to utilities in their decision-making processes for developing non-leaded component replacement programs.
- A research agenda related to the use of non-leaded materials. This report can be used as a framework for organizing and prioritizing research funding decisions. The research agenda encompasses a broad variety of issues related to non-leaded components.
- A summary of NSF/ANSI Standard 61 Sections 4, 8, and 9 procedures. This summary provides a useful overview of the evaluation process for products tested for certification according to this standard.
- A preliminary evaluation of the leaching potential of alternative brasses using NSF/ANSI Standard 61, Section 9 synthetic water and a modified water more representative of typical drinking water quality in the United States. This evaluation represents a first step towards developing research information that can be used to understand the performance of non-leaded materials and provide information with which to re-assess current regulatory and certification procedures.

CHAPTER 1
INTRODUCTION AND BACKGROUND

INTRODUCTION

The objective of this Awwa Research Foundation (AwwaRF) research effort was to provide definition and structure to the issues surrounding the current state of knowledge, testing protocols, performance, regulatory environment, and research gaps pertaining to the widespread use of utility owned non-leaded brass materials.

To achieve that goal, a review and summary of the current understanding of non-leaded brass plumbing and distribution components were completed, the major issues related to their manufacture, specification, and use in potable water distribution systems were identified, and a plan was prepared to address identified research data gaps and practical information related to their use. The desire to remove all lead from potable water distribution systems has driven the development of non-leaded brass alloys. Meters, components, and fittings manufactured from these non-leaded alloys are being increasingly installed and even specified for use in water distribution systems and premise plumbing. Despite the move to specify and install non-leaded materials, several unanswered questions remain regarding the long-term performance and the potential for metal release. This project identified current unknowns surrounding the use and performance of non-leaded brasses in meters, components and fittings and developed a framework for addressing those questions in a prioritized manner. The findings of this project have been developed based on input by industry experts, utility representatives, regulators, and manufacturers.

BACKGROUND

The following sections present background information on federal regulations governing drinking water and components used in drinking water systems and NSF International (NSF) definitions of drinking water components.

Federal Regulations

Prior to the promulgation of the Lead and Copper Rule (LCR) in 1991, lead pipe, lead solder and flux, and faucets and components composed of brass were shown to be major sources of lead in tap water. To control these sources, the 1986 amendments to the Safe Drinking Water Act (Section 1417(a)(1)) banned the use of lead pipe and required the use of 'lead-free' solders, fluxes, pipes, pipe fittings and well pumps in installation or repair of public water systems or plumbing that provided water from human consumption. Solders and fluxes may not contain more than 0.2 percent lead while lead-free pipes, pipe fittings, and well pumps may not contain more than 8.0 percent lead [U.S. Environmental Protection Agency (USEPA) 1987].

While lead solder has been banned, the installation of new components made of various copper alloys containing up to 8 percent lead was still allowed. While the objective of the lead-free products was the reduction of lead in potable water systems, the net result in a limited number of cases has been a fresh source of lead added that can contribute significant lead to potable water. Federal LCR requirements do not cover hose bibs, bathtub faucets, shower heads, or industrial faucets. The 1988 Lead Contamination Control Act (LCCA) focused on lead contamination of school drinking water supplies. The LCCA required that USEPA identify drinking water cooler brands with lead and non-lead liners, and required that all lead-lined coolers be replaced (USEPA1988).

The 1996 Safe Drinking Water Act (SDWA) Amendments required that USEPA establish a performance standard for lead leaching from fittings used for drinking water if voluntary standards were not implemented by the industry (USEPA 1996). In 1997, the USEPA determined that NSF/ANSI Standard 61, Section 9 satisfied the 1996 SDWA requirement for that voluntary standard (USEPA 1997). As a result, plumbing fittings that have a lead content of 8 percent or less and are NSF/ANSI Standard 61, Section 9 certified can be defined as "lead free". Utility in-line components are not certified under Standard 61, Section 9, but rather are certified under Standard 61, Section 8 procedures. Since Standard 61, Section 8 was not explicitly specified in the 1997 Federal Register notice as meeting the requirement for a voluntary standard, it appears that in-line utility components may not require certification, but must not contain more than 8 percent lead to be defined as 'lead free' and be installed in potable water systems.

Since early 2004, the USEPA has conducted a comprehensive national review of compliance and implementation issues associated with the LCR to determine if there is a national problem related to elevated lead levels. This review was undertaken primarily in response to lead action level exceedences experienced by the District of Columbia (DC) Water and Sewer Authority (DCWASA). The comprehensive review consisted of several elements, including a series of workshops examining the following specific issues:

- Simultaneous Compliance
- LCR Monitoring Protocols
- Public Education
- Lead Service Line Replacement
- Lead in Schools and Childcare Facilities
- Lead in Plumbing Fittings and Fixtures.

Workshop participants included topic experts from utilities, academia, state governments, and other stakeholder groups as well as observers with particular interest in the LCR impact on the drinking water community. Other elements of the comprehensive review included a data review to evaluate the effectiveness of the rule, and a review of LCR implementation by States and utilities. As a result of the comprehensive national review, USEPA identified opportunities to improve and clarify specific areas of the LCR and guidance materials to be addressed in a Drinking Water Lead Reduction Plan. One of several issues identified by the USEPA requiring longer-term consideration includes the revision to lead content in plumbing fittings requirements (USEPA 2005a).

In-Line Components/Plumbing Devices

In-line components used in the transmission and distribution of drinking water include mechanical devices such as valves, meters, backflow prevention devices, pressure regulators, and pump components, and connection devices such as fittings, couplings, meter setters, corporation stops, and curb stops. These components, or their parts, are generally manufactured using some type of copper alloy material that contains lead as an alloying element.

On the other hand, NSF/ANSI Standard 61, Section 8 contains a voluntary standard related to inline devices, components, and materials used in the transmission and distribution system (NSF 2003). It defines in-line devices as any device used to measure or control the flow of water that is installed on a service line or in a building distribution system downstream of the water main and before end point devices (i.e. faucets and other components). Examples of devices listed in this standard include:

- Backflow preventers
- Pressure regulators
- Compression fittings
- Strainers
- Expansion tanks
- Meter stops
- Check valves
- Service saddles
- Curb stops
- Water meters
- Building valves
- Pressure tanks
- Corporation stops
- Valves and fittings
- Meter couplings.

NSF/ANSI Standard 61, Section 9 specifies procedures for certifying plumbing devices, components and materials that are installed within the last liter of the distribution system (i.e. endpoint devices). Devices that are included in this include:

- Single and double handled faucets
- Hot and cold water dispensers
- Drinking fountains
- Bubblers and water coolers
- Residential refrigerator ice makers
- Supply stops and endpoint control valves
- Commercial kitchen devices.

Finally, NSF/ANSI Standard 61 Section 4 identifies certification procedures for pipes and related products. Products included in this section of the standard include:

- Fittings and couplings
- Flexible and rigid tubing
- Hoses
- Well casings
- Screens
- Pipe-related coatings.

Appendix A contains a summary of the evaluation procedures specified in these sections of NSF/ANSI Standard 61. ***It should be emphasized that the primary focus for this project is to identify the issues surrounding the manufacture and use of non-leaded, utility controlled in-line components, rather than customer owned and installed faucets. However, the same materials are often used in both 'in-line' and 'end-use' components so both are discussed in this project report.***

Definition of "Lead Free", 'Low Lead", and "Non-Lead

There are a variety of terms used to describe the lead content of materials and components. Leaded materials are those solders, fluxes, brasses, pipes, and fittings typically installed prior to their 1986 ban. These materials contain a varying amount of lead as a percent of their total composition. The 1986 Safe Drinking Water Act amendments banned solder and flux containing more than 0.2 percent lead and any pipe and plumbing fitting or fixture that was not "*lead-free*." That legislation defined "*lead-free*" as a material having a lead content of 8 percent or less of total content. While the regulatory term is "*lead-free*", these components can still contain up to 8 percent lead. New non-leaded brass alloys contain between 0.1 to 0.25 percent lead. No lead is added to these alloys, and the minimal amount of lead present is due to contributions from recycled materials and ore that are used as the source metals. These alloys, and components manufactured using these alloys, will be referred to as "*non-lead*" or "*non-leaded*" in this report. Alloys and components containing 8 percent lead or less (including non-leaded alloys) have generally been referred to as "*low-lead*'.

For purposes of this report, the term "*lead free*" will be used to identify those materials and components that meet the requirements of the 1986 Safe Drinking Water Act amendments, i.e. having a lead content of 8 percent or less. "*Non-Lead*" or "*non-leaded*" will be used to identify those materials and components that are made from alloys where no additional lead has been added. The term "*low lead*" will not be used in this report.

CHAPTER 2
CURRENT UNDERSTANDING OF NON-LEADED BRASS

INTRODUCTION

In order to provide a current understanding of the issues surrounding the use of non-leaded, residential-sized components in the water distribution system and premise plumbing, a series of papers were developed on the following topics:

- Description of the Composition, Lead Content and Mechanical Properties of New Non-Leaded Brasses
- Potential Health Effects of Non-Leaded Brasses
- Potential of Non-Leaded Brasses to Release Lead, Copper, and other Metals (Bismuth and Selenium) into Drinking Water
- American Society for Testing and Materials (ASTM) Specifications for New Non-Leaded Brasses
- Summary of Current State Experiences with Non-Leaded Materials
- Utility Motivations and Experience with Non-Leaded Brass Components
- Potential Impacts to Manufacturers
- Plumbing Industry Issues
- Potential Impacts to Utility Operations and Maintenance Procedures
- Alternative Materials for Fittings and Components.

The purpose of these topic papers was to formulate an initial understanding of the issues surrounding the use of non-leaded, residential-sized components in the water distribution system and premises plumbing in order to provide context to the Expert Workshop. In addition, these topic papers summarize the current knowledge on the performance of non-leaded brass alloys and their potential for leaching compounds (lead, copper, selenium, bismuth) in drinking water. This information is also of particular value to utilities in their decision-making processes for developing non-leaded component replacement programs. These topic papers are included below.

DESCRIPTION OF THE COMPOSITION, LEAD CONTENT, AND MECHANICAL PROPERTIES OF NEW NON-LEADED BRASSES

Research has been undertaken in order to develop a new generation of reasonably priced alloys for plumbing hardware that contain no added lead ($\leq$ 0.25 percent). Historically, lead has been added to brass alloys used in drinking water applications to improve the "pressure-tightness" of a product by sealing micro-pores (Showman 1994). Lead is also believed to reduce the corrosion of other metals in the alloy and result in longer product life (OECD 1994). In addition, lead is important in the machining of brass castings. It functions both as a "chip breaker" and a lubricant, resulting in cost-effective machining rates for intricate brass devices (Showman 1994).

The presence of lead in the alloy also increases the potential for contamination of water that contacts the device. Under at least some circumstances, the water flowing through the device can dissolve the "pockets of lead" that exist on the brass surface, and introduce lead into drinking water (Cassidy 1994). Elevated lead levels in drinking water have been attributed to lead containing products. For instance, in a survey that was conducted by the Association of Metropolitan Water Agencies for the promulgation of the LCR, lead leaching from new faucets and fittings was found to be, or suspected to be, the major reason for increased lead levels in water samples (Lytle and Schock 1996). Other researchers have drawn very strong circumstantial ties between lead in water and leaching from brass (Kimbrough 2001).

Composition

Currently, three families of non-leaded brass alloys are commercially available:

- The combined effort of the CDA, the Brass and Bronze Ingot Manufacturers (BBIM), the American Foundrymen's Society (AFS), the Materials Technology Laboratory of CANMET in Canada, other foundries and water product manufacturers resulted in a series of non-leaded copper alloys available as Envirobrass® (SeBiLOY®). These alloys contain from 0.1 percent lead (Envirobrass® III) to 0.25 percent lead (Envirobrass® I and II), as an incidental impurity from the recycling of materials from which casting alloys are commonly made (i.e. no lead is intentionally added to them) (CDA 2000). In the place of lead, bismuth and selenium are added to the alloys to improve mechanical characteristics.
- The Federal Metal Company has also developed non-leaded alloys commercially available as FederAlloy®. In these alloys, bismuth is added in the place of lead, which comprises up to 0.1 percent of the alloy in most cases (Table 2.1), due to incidental impurities. No selenium is added to the FederAlloy® series of brasses.
- Sambo Copper Alloy Co., Ltd developed a silicon brass with the trade name ECO BRASS®. The alloy was originally designed in Japan to meet the Japanese 7 μg/L lead leachate requirement for copper alloy drinking water components. The alloy contains silicon and phosphorus (Table 2.1) in a brass with a nominal 20 percent zinc content. There are two subsets of this alloy designed for sand and permanent mold casting, with each having a lead maximum of 0.1 percent to accommodate contaminants from scrap used as charge materials.

The compositions of Envirobrass®, FederAlloy® and ECO BRASS® alternative brasses (Table 2.1) were developed so that products made from these alloys would be comparable to their leaded counterparts in terms of mechanical strength.

Mechanical Strength Properties

Mechanical properties such as ultimate tensile strength, yield strength, elongation and hardness have been determined for non-leaded brass alloys, in order to evaluate their strength. The mechanical properties of Envirobrass® I and Envirobrass® II are in general similar to the standard leaded semi-red and red plumbing brasses. However, they are lower in elongation. Table 2.2 presents a summary of mechanical strength properties for representative leaded and non-leaded alloys. Envirobrass® III strength properties were not all directly comparable to those of leaded yellow brass (Table 2.2). For FederAlloy® brasses I and III tensile and yield strength, elongation and hardness approximate their leaded alloy equivalents. ECO BRASS® sand and permanent mold castings (C87851 and C87852 respectively) in general have higher tensile strength, yield strength and hardness when compared to the conventional leaded alloys (Table 2.2).

Machinability

Machining is used as the final shaping operation in all plumbing cast products and contributes significantly to the cost (CDA 2005). In assessing the machinability of an alloy, factors such as tool life, surface quality, chip form and cutting force are considered (Kuyucak and Sahoo 1996). The leaded alloy C36000 (free-cutting brass) has the maximum machinability of 100. Compared to this machinability rating, copper alloys are generally classified into three categories (Kuyucak and Sahoo 1996):

- Free-machining alloys, which have ratings between 70 - 100
- Readily machinable alloys, which have ratings between 30 - <70
- Difficult-to-machine alloys, which have ratings less than 30.

Based on the above classification Envirobrass® I, II and III alloys all fall within the free-machining range of 70-100. From the FederAlloy® family of alloys, and for those for which information was readily available, all types but one fall within the free-machining range. The exception is the (original) FederAlloy® I C89837, with a slightly lower machinability of 66. The alternative alloy ECO BRASS® falls within the free-machining range; it has a machinability of 70, at the lower end of that range (Table 2.2).

Table 2.1
Composition of some current non-leaded brass alloys and commonly used leaded alloys for comparison. (CDA 2005 and Federal Metal Company 2005)

	ALLOY	RANGE OR MAXIMUM %													
		Copper (*Cu*)	Tin (*Sn*)	Lead (*Pb*)	Zinc (*Sn*)	Bismuth (*Bi*)	Selenium (*Se*)	Nickel (*Ni*)	Iron (*Fe*)	Antimony (*Sb*)	Sulfur (*S*)	Phosphorous (*P*)	Aluminum (*Al*)	Silicon (*Si*)	Manganese (*Mn*)
ENVIRO BRASS®	Envirobrass® I (C 89510)	86.0-88.0	4.0-6.0	0.25	4.0-6.0	0.5-1.5	0.35-0.75	1.0	0.2	0.25	0.08	0.05	0.005	0.005	-
	Envirobrass II (C89520)	85.0-87.0	5.0-6.0	0.25	4.0-6.0	1.6-2.2	0.8-1.1	1.0	0.2	0.25	0.08	0.05	0.005	0.005	-
	Envirobrass® III (C89550)	58.0-64.0	1.2	0.1	32.0-38.0	0.6-1.2	0.01-0.1	1.0	0.5	0.05	0.05	0.01	0.1-0.6	0.25	-
FEDERALLOY ENVIRO FEDERALLOY®	FederAlloy® I original (C89837)	84.0-88.0	3.0-4.0	0.10	6.0-10.0	0.70-1.2	-	0.50-1.0	0.30	0.25	0.08	0.05	0.005	0.005	-
	FederAlloy® I - 836 Version (C89833)	87.0-91.0	4.0-6.0	0.10	2.0-4.0	1.7-2.7	-	0.50-1.0	0.30	0.25	0.08	0.05	0.005	0.005	-
	FederAlloy® I - 836 Version Zero Lead	87.0-91.0	4.0-6.0	0.005	2.0-4.0	1.7-2.7	-	0.50-1.0	0.30	0.25	0.08	0.05	0.005	0.005	-
	FederAlloy® I - 833 Version	92.0-94.0	1.0-2.0	0.10	2.0-6.0	1.2-1.7	-	0.1	0.10	Trace	Trace	0.05	0.005	0.005	-
	FederAlloy® I - 844 Version (C89831)	87.0-91.0	2.7-3.7	0.10	2.0-4.0	2.7-3.7	-	0.50-1.0	0.30	0.25	0.08	0.05	0.005	0.005	-
	FederAlloy® I - 844 Modified (C89844)	83.0-86.0	3.0-5.0	0.20	7.0-10.0	2.0-4.0	-	1.0	0.30	0.25	0.08	0.05	0.005	0.005	-
	FederAlloy® I 848-A Version	77.0-79.0	3.0-4.0	0.10	15.5-17.5	1.7-2.7	-	0.50-1.0	0.35	0.25	0.08	0.05	0.005	0.005	-
	FederAlloy® II (922 Version)	Balance	5.75-6.50	0.10	3.0-5.0	1.0-1.75	-	0.50-1.0	0.20	0.25	0.05	0.05	0.005	0.005	-
	FederAlloy® II (927 Version)	Balance	9.0-11.0	0.10	0.7	1.0-2.0	-	0.50-1.0	0.20	0.25	0.05	0.05	0.005	0.005	-
	FederAlloy® II (255 Ingot Version)	Balance	5.0-6.0	0.10	3.5-4.5	1.7-2.7	-	0.80	0.25	0.20	0.05	0.010-0.016	0.005	0.005	-

(continued)

Table 2.1 (Continued)
Composition of some current non-leaded brass alloys and commonly used leaded alloys for comparison.
(CDA 2005 and Federal Metal Company 2005)

ALLOY	RANGE OR MAXIMUM %													
	Copper (Cu)	Tin (Sn)	Lead (Pb)	Zinc (Sn)	Bismuth (Bi)	Selenium (Se)	Nickel (Ni)	Iron (Fe)	Antimony (Sb)	Sulfur (S)	Phosphorous (P)	Aluminum (Al)	Silicon (Si)	Manganese (Mn)
FederAlloy® III (932 Version) (C89835)	85.0-89.0	6.0-7.5	0.10	2.0-4.0	1.7-2.7	-	0.50-1.0	0.20	0.25	0.08	0.05	0.05	0.005	-
FederAlloy® III (937 Version) (C89325)	84.0-88.0	9.0-11.0	0.10	1.0	2.7-3.7	-	0.50-1.0	0.15	0.25	0.08	0.05	0.05	0.005	-
FederAlloy® III (938 Version)	Balance	6.0-8.0	0.10	1.0	3.7-4.7	-	0.50-1.0	0.15	0.50	0.08	0.05	0.05	0.005	-
FederAlloy® III (945 Version)	86.0-90.0	6.0-8.0	0.10	1.2	3.7-4.7	-	0.50-1.0	0.15	0.25	0.08	0.05	0.05	0.005	-
FederAlloy® IV (B1 Version) Sand Cast Version	60.0-63.0	0.50-1.50	0.10	Balance	0.75-1.5	-	1.0	0.50	0.02-0.10	0.05	0.01	0.50-0.75	0.05	-
FederAlloy® IV (B2 Version) Permanent Mold Version	59.5-60.5	0.10	0.10	Balance	0.75-1.5	-	0.01	0.03	0.01	0.01	0.005	0.50-0.75	0.02	-
FederAlloy® V (973 Version)	63.0-67.0	1.50-3.0	0.10	13.00-17.00	1.0-2.0	-	11.00-14.00	0.75	Trace	0.08	Trace	0.005	0.005	0.5
FederAlloy® V (974 Version)	58.0-61.0	2.50-4.50	0.10	Balance	0.75-1.0	-	16.00-17.00	1.5	0.25	0.08	0.02	0.005	0.05 0.015	0.5
FederAlloy® V (976 Version)	Balance	3.50-4.50	0.10	3.0-9.0	0.75-1.0	-	19.00-21.50	1.5	0.25	0.08	0.05	0.005		1.0

(continued)

Table 2.1 (Continued)
Composition of some current non-leaded brass alloys and commonly used leaded alloys for comparison.
(CDA 2005 and Federal Metal Company 2005)

ALLOY		RANGE OR MAXIMUM %													
		Copper (Cu)	Tin (Sn)	Lead (Pb)	Zinc (Sn)	Bismuth (Bi)	Selenium (Se)	Nickel (Ni)	Iron (Fe)	Antimony (Sb)	Sulfur (S)	Phosphorous (P)	Aluminum (Al)	Silicon (Si)	Manganese (Mn)
ECO BRASS®	ECO BRASS® (C87850)	74.0-78.0	0.20	0.10	Balance	-	-	0.10	0.10	-	-	0.04-0.20	-	2.7-3.4	0.10
LEADED	C83600 (Red Brass)	84.0–86.0	4.0-6.0	4.0-6.0	4.0-6.0	-	-	1.0	0.3	0.25	0.08	0.05	0.005	0.005	-
	C84400 (Semi Red Brass)	78.0–82.0	2.3-3.5	6.0-8.0	7.0-10.0	-	-	1.0	0.4	0.25	0.08	0.02	0.005	0.005	-
	C85800* (Yellow Brass)	57.0 (min)	1.5	1.5	31.0-41.0	-	-	0.5	0.5	0.05	0.05	0.01	0.55	0.25	0.25

***Die Cast**

Table 2.2
Typical mechanical properties and machinability rating for some current non-leaded brasses and for commonly used traditional leaded brasses. The alloys were sand cast unless otherwise stated. (CDA 2000 and Chase Brass and Copper Co. 2006)

ALLOY		UTS-Ultimate Tensile Strength ksi (Mpa)	0.5 % YS-Yield Strength ksi (Mpa)	Elongation (%)	Hardness (500 kg) BHN†	Machinability Rating (-)
LEADED	C83600 (Red Brass)	37 (255)	17 (117)	30	60	84
	C84400 (Semi Red Brass)	34 (234)	15 (103)	26	55	90
	C85800* (Yellow Brass)	55 (379)	30 (207) (but at 0.2% offset)	15	60 (but as Rockwell hardness B)	80
ENVIRO BRASS®	C89510 (Envirobrass® I)	30 (209)	20 (136)	12	37	75
	C89520 (Envirobrass® II)	31 (215)	21 (145)	10	54	85
	C89550 (Envirobrass® III)	48 (330)	29 (200)	8	-	75 (with 0.7% Bi) 97 (with 1.1% Bi)
FEDERALLOY®	FederAlloy® I (C89837)	37 (258)	17 (119)	32	60	66
	FederAlloy® I-836 (C89833)	37 (258)	17 (119)	28	60	81
	FederAlloy® I-844 (C89831)	34 (237)	15 (105)	20	55	85
	FederAlloy® I-844 Modified (C89844)	34 (234)	15 (103)	30	55	70
	FederAlloy® III-932 (C89835)	35 (244)	18 (126)	20	65	70
	FederAlloy® III-937 (C89325)	35 (244)	18 (126)	20	73	80
ECO BRASS®	ECO BRASS® C87851	60 (379)	24 (165)	22	115	70
	ECO BRASS® C87852	69 (475)	35 (241)	21	135	70

*Die Cast

† BHN= Brinnel Hardness Number

POTENTIAL HEALTH EFFECTS OF NON-LEADED BRASS ALLOYS

Non-leaded brasses can contain a variety of metals that may be released into the drinking water, including bismuth, selenium, copper, tin, nickel, antimony, zinc, and lead. The following sections summarize information on the potential health effects of these metals in drinking water.

Bismuth

Bismuth, a reddish white metal, is typically found in nature as bismite (Bi_2O_3), bismuthinite (Bi_2S_3), and bismutite [$(BiO)_2CO_3$]. It occurs in two oxidation states, +3 and +5. Most health information on bismuth has focused on pharmaceutical applications (USEPA 2005b), probably due to the fact that the main applications of bismuth salts (such as bismuth subsalicylate, the active ingredient in Pepto Bismol™ and Kaopectate™) are for the treatment of gastrointestinal problems like gastritis, peptic ulcer and diarrhea. The World Health Organization (WHO 2005) states that "acute poisoning with bismuth may cause renal damage, encephalopathy and peripheral neuropathy. Neurotoxicity may also occur after chronic treatment with bismuth salts, but is reversible if medication is stopped." The USEPA is currently reviewing the health effects of bismuth to determine if a health advisory is warranted (USEPA 2005b), and NSF International is also conducting a risk assessment on bismuth (McClellan 2005). As part of the NSF International risk assessment, analytical tests will be performed to determine the form of Bi leached from brass, specifically its oxidation state and whether it is an oxide of chloride salt. These are being conducted primarily to confirm that Bi^{3+} is present as an oxide. There is also an issues related to the Bi form in the colloids that have been used in available toxicological studies. There is a need to establish the chemical relationship between the form of Bi from brass and the form of Bi in the colloids as well as the expected absorption of Bi from brass compared to the colloids (McClellan 2006). Bismuth is not currently regulated in drinking water and its occurrence in drinking water has not been documented.

Selenium

Selenium is a naturally occurring mineral element that is distributed widely in nature in most rocks and soils. In its pure form, it exists as metallic gray to black hexagonal crystals, but in nature it is usually combined with sulfide or with silver, copper, lead, and nickel minerals. Selenium is an essential nutrient at low levels. Recommended daily allowance for men and women is 0.055 mg/d (Agency for Toxic Substances and Disease Registry 2003a). ATSDR and the EPA have determined that 5 micrograms of selenium per kilogram of body weight taken daily would not be expected to cause any adverse health effects over a lifetime of such intake.

The USEPA has set a Maximum Contaminant Level Goal (MCLG) and Maximum Contaminant Level MCL of 0.05 mg/L for selenium because it can potentially cause the following health effects at levels above the MCL: hair and fingernail changes, damage to the peripheral nervous system, fatigue and irritability, damage to kidney and liver tissue, as well as the nervous and circulatory systems. Currently, there is no evidence that selenium has the potential to cause cancer from lifetime exposures in drinking water. In fact, based on human studies, lower-than-normal selenium levels in the diet might increase the risk of cancer (Agency for Toxic Substances and Disease Registry 2003b). Selenium is usually found at very low levels in drinking water (<0.01 mg/L). Higher levels may occur in drinking waters in areas where high selenium levels in the soil contribute to elevated levels in the water (USEPA 2005c).

Copper

Copper is a reddish metal that occurs naturally throughout the environment, in rocks, soil, water, and air. Copper is an essential nutrient at low levels, however toxic effects can occur at higher levels. Acute exposure can cause gastrointestinal disturbances, including nausea, vomiting, stomach cramps, or diarrhea, and intentionally high intakes of copper can cause liver and kidney damage and even death. Chronic exposure can cause liver or kidney damage and people with Wilson's disease may be more sensitive to the effects of copper contamination. USEPA does not classify copper as a human carcinogen because there are no adequate human or animal cancer studies." (Agency for Toxic Substances and Disease Registry 2004 and USEPA 2005d).

Copper levels in drinking water generally occur due to corrosion of copper premise piping. USEPA has set the MCLG and an action level of 1.3 mg/L and a secondary drinking water standard of 1 mg/L. The MCLG for copper was set based on the potential for health effects at levels greater than 1.3 mg/L, and the action level was set based on the ability of present treatment technology to control copper levels at customers' taps. The secondary standard was established because of blue-green staining and metallic taste associated with copper levels above 1 mg/L. These secondary standards are federally non-enforceable guidelines, however some states (such as California) may have state statutes in place to enforce secondary standards.

Tin

Tin is a soft, white, silvery metal that is insoluble in water. Tin can combine with other chemicals such as chlorine, sulfur or oxygen to form inorganic tin compounds. Tin can also combine with carbon to form organotin compounds. Tin metal, as well as inorganic and organic tin compounds can be present in water. Organic tin compounds are mostly attached to particles in water or may settle out into sediments.

Because inorganic tin compounds usually enter and leave the body rapidly after ingestion, they do not usually cause harmful effects. However, humans who swallowed large amounts of inorganic tin in a research study suffered stomach aches, anemia, and liver and kidney problems. There is no evidence that inorganic tin compounds affect reproductive functions, produce birth defects, or cause genetic changes. Exposure to some organotin compounds has been shown to cause harmful effects in humans, but the main effect depends on the particular compound. There have been reports of skin and eye irritation, respiratory irritation, gastrointestinal effects, and neurological problems in humans exposed for a short period of time to high amounts of certain organotin compounds." (Agency for Toxic Substances and Disease Registry 2005a). There is no evidence that tin or tin compounds cause cancer in humans. There are no U.S. regulations for tin in drinking water.

Zinc

Zinc is one of the most common elements in the earth's crust. Zinc is found in the air, soil, and water and is present in all foods. "Zinc is an essential element needed by your body in small amounts. We are exposed to zinc compounds in food and it is also present in most drinking water. Drinking water or other beverages may contain high levels of zinc if they are stored in metal containers or flow through pipes that have been coated with zinc to resist rust." (Agency for Toxic Substances and Disease Registry 2003c)

The USEPA has not established a primary (health based) standard for zinc in drinking water. However, a secondary standard of 5 mg/L zinc has been established because of effects on the water's taste (metallic). Zinc has not been classified for carcinogenicity due to incomplete information. (Agency for Toxic Substances and Disease Registry 2003d).

Nickel

Nickel is a silvery white metal that is hard, malleable, and ductile. It is found in natural ore deposits with other elements and is primarily used to make stainless steel and other alloys. "Eating or drinking large amounts of nickel has caused lung disease in dogs and rats and has affected the stomach, blood, liver, kidneys, and immune system in rats and mice, as well as their reproduction and development. However, exposures of the general population from other sources, such as foods and drinking water, are almost always too low to be of concern" (Agency for Toxic Substances and Disease Registry 2003e).

The USEPA promulgated an MCLG and MCL for nickel of 0.1mg/L. These were based on health effects associated with chronic exposure to levels above the MCL, which include decreased body weight, heart and liver damage, and skin irritation. However these regulations were remanded on February 9, 1995 and there is no current regulation for nickel in drinking water (USEPA 2005e).

Antimony

Antimony is a silvery-white metal that is found in the earth's crust. Antimony is not used alone because it breaks easily, but when mixed into alloys, it is used in lead storage batteries, solder, sheet and pipe metal, bearings, castings, and pewter (Agency for Toxic Substances and Disease Registry 1995). Because antimony is found naturally in the environment, the general population is exposed to low levels of it every day, primarily in food, drinking water, and air.

The USEPA has established an MCLG and MCL for antimony of 0.006 mg/L in drinking water because of health effects including nausea, vomiting and diarrhea. The Department of Health and Human Services (DHHS), the International Agency for Research on Cancer (IARC), and the USEPA have not classified antimony as to its human carcinogenicity (Agency for Toxic Substances and Disease Registry 1995).

Lead

Lead is a heavy, bluish-gray metal that occurs naturally in the Earth's crust, usually associated with ores containing other elements. Because of its physical properties (corrosion resistance, malleability) lead and lead alloys have historically been used in plumbing applications. Elevated lead levels in drinking water generally occur due to corrosion of lead pipe used for service lines and lead contained in household plumbing. A variety of health effects have been associated with lead exposure, including interference with red blood cell chemistry, delays in normal physical and mental development in babies and young children, slight deficits in the attention span, hearing, and learning abilities of children, and slight increases in the blood pressure of some adults. Long-term effects of lead exposure include stroke and kidney disease, and cancer (USEPA 2005f). Exposure to lead, especially in infants and young children, may cause permanent neurological damage leading to IQ reductions, attention deficit disorders, and aggressive behavior (Needleman and Gatsonis 1990; Schwartz 1994). Too much lead can also damage reproductive systems and the kidneys, and can cause high blood pressure and anemia (WHO 1993).

Kidney tumors have developed in rats and mice that had been given large doses of some kinds of lead compounds. The Department of Health and Human Services (DHHS) has determined that lead and lead compounds are reasonably anticipated to be human carcinogens based on limited evidence from studies in humans and sufficient evidence from animal studies, and the USEPA has determined that lead is a probable human carcinogen. The International Agency for Research on Cancer (IARC) has determined that organic lead compounds are "not classifiable as to their carcinogenicity to humans (Group 3) based on inadequate evidence from studies in humans and in animals." IARC has classified inorganic lead compounds as probable human carcinogens (Group 2A). (ATSDR 2005b, IARC 2006).

In light of these findings, the USEPA set a MCLG of zero for lead. The LCR lead action level of 0.015 mg/L at the 90th percentile is not a health-based action level but was derived based on the capabilities of corrosion control processes to adequately control lead concentrations at the tap (USEPA 2004a). The WHO has established a health guideline for lead in drinking water at 0.1 mg/L (WHO 1993). The US Centers for Disease Control and Prevention has set a level of 10 μg of lead per deciliter (dL) of blood as the level above which a child is to be investigated for lead poisoning. The USCDC states that efforts to eliminate lead exposures through primary prevention have the greatest potential for success in reducing blood lead levels (USCDC 2005).

Recent published research has documented measurable childhood IQ deficits at a blood lead level as low as 2.5 μg/dL, a level previously believed to be well below the threshold for observable neurological damage (Hertz-Picciotto et al. 2000; Lanphear et al 2000; Bellinger 2003). Canfield et al (2003) evaluated the relationship between blood lead levels and IQ in children using linear and nonlinear models and concluded that children in the U.S. may be more adversely affected by environmental lead than previously thought. Their results indicated that each 10 μg/dL increase in the lifetime average blood lead concentration was associated with a 4.6-point decrease in IQ. However, for a sub-sample of children with blood lead levels below 10 μg, the change in IQ associated with a change in blood lead concentration from 1 to 10 μg /dL was greater (7.4 points).

Very recent published research (Menke et al. 2006) demonstrated a link between blood lead levels below 10 μg/dL, and death by heart attack and stroke. The effects begin to appear at ~2 μg/dL, or one-fifth of the current USCDC blood lead intervention level. This relationship was previously unknown, and if confirmed, is likely to lead to a reduction in the USCDC blood lead intervention level. The Menke et al study resulted in an editorial in the journal in which it appeared (Nawrot and Staessen 2006), calling attention to the significance of the demonstrated effects. A proposal to reduce the current USCDC blood lead intervention level from 10 to 2 μg/dL has also been published (Gilbert and Weiss 2006). The change is justified on the basis of health effects research and present laboratory accuracy for blood lead determinations. Any reduction in the USCDC blood lead intervention level is likely to cause the USEPA to reexamine the Action Level for lead in drinking water.

POTENTIAL OF NON-LEADED BRASSES TO RELEASE LEAD AND COPPER AND OTHER METALS (BISMUTH AND SELENIUM) INTO DRINKING WATER

The potential of non-leaded brasses to leach lead, copper and other metals into potable water has not been adequately examined. Preliminary tests studying the corrosion and dezincification behavior of non-leaded brasses have been aimed more at their practical performance in drinking water applications rather than their potential health effects.

More recently, and as the discussion on non-lead brass alternatives is gaining further attention, research studies were conducted that specifically addressed the leaching issues. However, it is obvious at this point that more work needs to be undertaken in order to gain a better understanding of the leaching propensity of these alloys.

Leaching Studies

Among the few studies that have been conducted, the American Foundry Society (AFS) quantified leaching levels for two different experimental non-lead compositions in brasses (Table 2.3). Hollow cylinders 5'' long x ¼'' diameter were tested in the "as-cast" form. One sample of C84400 served as a control. Initial results were obtained at two pH levels of 5 and pH 10, purportedly following the NSF/ANSI Standard 61 Section 9 methodology (Table 2.4). There may be an editorial or other error in this report, since Section 9 does not use a pH 5 and pH 10 water. It is also uncertain at this time whether the pH 5 water and the pH 10 water had orthophosphate or were prepared fresh, which are two factors that can impact their aggressiveness towards lead. Despite these potential limitations, the results indicated that a 0.2 percent lead level in the alloy would be below the 11 μg/L maximum allowable level for the two test waters according to the Section 9 protocol. The leach levels for bismuth and selenium were also obtained at two different pH's. Bismuth leaching was below the detection limit for every experimental condition and at both pH levels (Table 2.5). The selenium leach level was below the detection limit for every experimental composition at pH = 5, but there were detectable levels of selenium leaching in some of the experimental alloys at pH = 10 (Table 2.5) (Twarog 1995).

The verification leach test that followed as part of the study was conducted with a pH = 8 solution, so that results could be comparable to other leaching tests conducted at pH = 8 using the NSF/ANSI Standard 61 Section 9 methodology. The alloy control that was used in this run was C83600. Tube samples were cut and adjusted so that they could maintain a 50 mL water volume. Before the test, all samples were washed and conditioned as described in Section 9 of NSF/ANSI Standard 61 (Appendix A). The "Q" statistic in the case of lead and the 19th day normalized mean concentration for bismuth and selenium were reported (Table 2.6) (Twarog 1995). Lead leaching was reduced by a factor of about 3-5 for the non-lead alloys as compared to the C83600 control.

Table 2.3.
Intended compositions of the two experimental non-lead brass alloys that were tested in the Twarog study (1995).

Experimental Alloy	Bi	Se	Sn	Zn	P	Ni	Pb	Cu
2%Bi/ 1% Se	2%	1%	5%	5%	0.15%	0.75%	0.2%	Bal.
1% Bi/ 0.5% Se	1%	0.5%	5%	5%	0.15%	0.75%	0.2%	Bal.

The compositions actually achieved when the multiple samples were fabricated, slightly deviated from the target values.

Table 2.4
Lead Leaching for Experimental Alloys (Twarog 1995).

Sample #	% Lead in Alloy	"Q" at pH=5 (ug)	"Q" at pH=10 (ug)
1	0.0	0.062	0.082
2	0.2	0.240	4.13
3	0.2	0.036	0.367
4	0.13	0.436	2.76
5	0.24	0.027	2.82
6	0.0	0.039	0.049
7	0.23	0.552	5.58
8	0.0	0.066	0.034
9	0.13	0.013	0.510
Control (C84400)	7.15	4.09	8.15

Table 2.5
19th day Mean Normalized Leach of Selenium and Bismuth for the Experimental Alloys.*
(Twarog 1995)

Sample Number	Bismuth pH=5	Bismuth pH=10	Selenium pH=5	Selenium (μg/L) pH=10
1	BDL	BDL	BDL	BDL
2	BDL	BDL	BDL	1.01 (μg/L)
3	BDL	BDL	BDL	0.23 (μg/L)
4	BDL	BDL	BDL	BDL
5	BDL	BDL	BDL	26.97 (μg/L)
6	BDL	BDL	BDL	8.05 (μg/L)
7	BDL	BDL	BDL	BDL
8	BDL	BDL	BDL	0.95 (μg/L)
9	BDL	BDL	BDL	54.56 (μg/L)
Control (C84400)	BDL	BDL	BDL	BDL

* BDL stands for Below Detection Limit levels

Table 2.6
Leach Verification Test Results in the Twarog Study (1995).

Alloy	Lead (μg/L) "Q" Statistic	Bismuth 19th Day Mean Normalized (μg/L)	Selenium 19th Day Mean Normalized (μg/L)
Control (C83600)	15.5	<0.1	<0.1
2%Bi/ 1%Se	5.6	1.6	1.7
1% Bi/ 0.5% Se	2.9	5.2	0.1

Other projects were undertaken on behalf of the CDA, which examined the leaching propensity of experimental alloys containing very small amounts of lead and bismuth (Peters 1995). Five brass alloys, including four alternative experimental alloys and the conventional leaded brass C84400 (Table 2.7), were tested according to an earlier version of NSF/ANSI Standard 61, Section 9. The test specimens were in the form of cast spouts. Based on the results of that study, the standard alloy C84400, with the highest lead content, resulted in the highest lead leaching (Figure 2.1). Alloy 2, with the lowest lead content of 1 percent, resulted in the lowest lead leaching, whereas alloys 3, 4 and 5 with the intermediate lead contents fell in between (Figure 2.1) (Peters, 1995).

Table 2.7
Compositions of experimental alloys reported in the CDA study (Peters, 1995).

Alloy #	Composition (% wt)					
	Pb	Bi	Sn	Zn	P	Cu
1 (C84440)	6.9	0	2.6	9.5	0.012	79.8
2	1	3	3	11.9	0.1	80.3
3	2	0.65	3	13.3	0.1	80.3
4	3	1	3	11.9	0.1	80.3
5	4	1.3	3	9.8	0.1	80.3

Subsequent experiments reported in the same CDA study, examined bismuth and selenium leaching from experimental alloys with very low lead content. The test specimens were large faucet body castings. The bismuth and selenium leaching data that were reported (Figure 2.2) were not adjusted for the internal volume of the test specimens. According to the report, when the 19th day concentrations are adjusted for the internal volume, the USEPA requirement for selenium (50 μg/L) is met. (Peters 1995).

In a 1992 study performed by the International Copper Development Association (ICA) and the CDA, the corrosion behavior of alloys containing 0.5, 1.0, and 1.5 percent lead was examined and compared to a 5 percent leaded alloy. The test method used in this study was based on a 1988 draft version of NSF/ANSI Standard 61, Section 12.0. This protocol is different that the one given in the current draft of the standard, therefore the data are not directly applicable to that test, but results are likely to give useful trends. The alloys were exposed to three potable waters at pH 5, 8 and 10 respectively, for 14 days at room temperature. Extracts from solutions exposed for 1, 2, 3, 4, 7, 8, 9, 10, 11 and 14 days, plus one final extraction were analyzed for lead using atomic absorption spectrometry. The effect of alloy lead content on lead leaching after a 14-day exposure period was reported for the pH 8 water (Figure 2.3). (Dresher and Peters 1992).

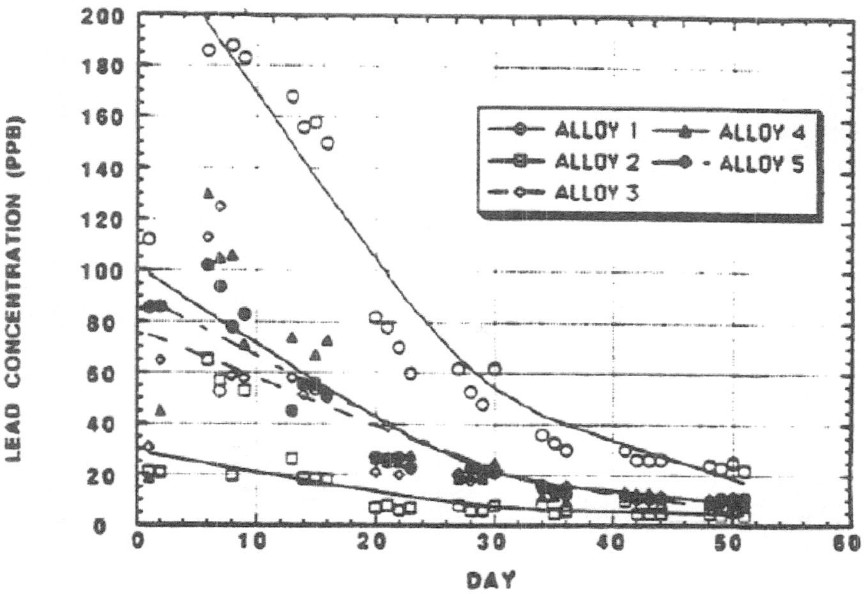

Figure 2.1. Lead leaching versus time in a pH 8.0 water, based on an earlier version of NSF/ANSI Standard 61, Section 9. (after Peters 1995)

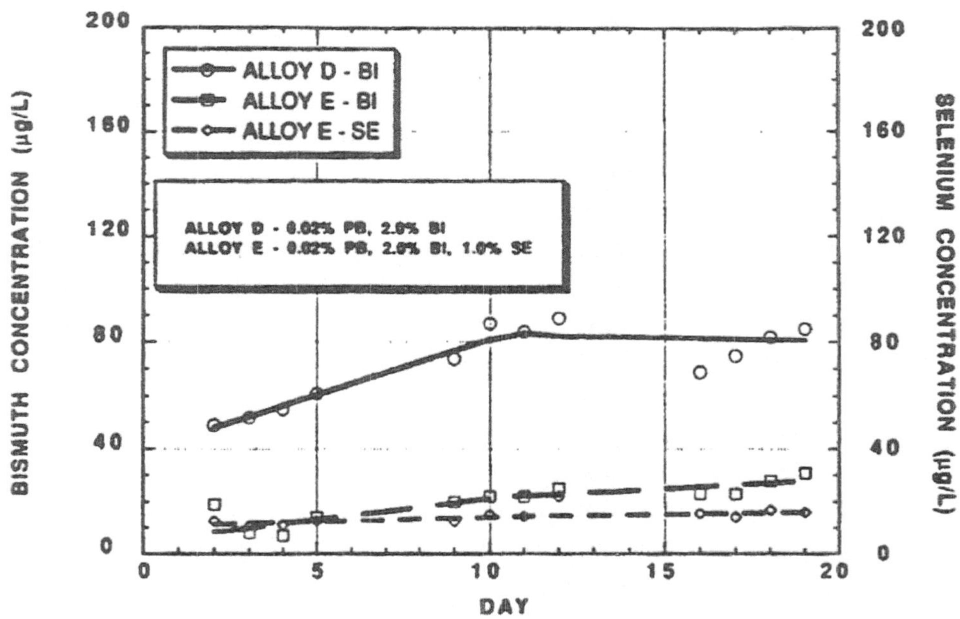

Figure 2.2. Bismuth and Selenium leaching in faucet body castings. (after Peters 1995)

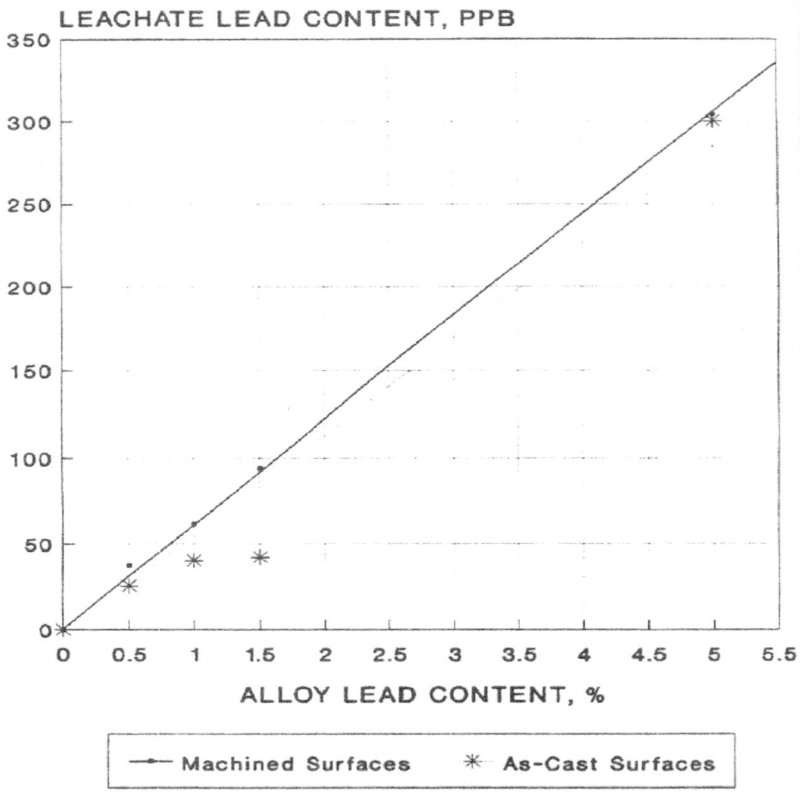

Figure 2.3. Effect of alloy lead content on lead leaching. Data were collected after a 14-day exposure in pH 8 water based on a 1988 draft of NSF/ANSI Standard 61, Section 12. This test protocol is no longer in use. (after Dresher and Peters 1992)

The Environmental Quality Institute performed laboratory studies in order to compare the leaching tendency of leaded (5-7 percent lead) versus non-leaded (<0.1 percent lead) individual plumbing components including curb stops, main stops, couplings and others. The components were exposed to a synthesized extraction water that closely matched the characteristics of average California public water supply. Although the parts were plumbed into a flow-through pilot apparatus and not "dumped and filled", NSF/ANSI Standard 61 Section 9 protocol was otherwise followed. Total lead was quantified at time intervals mandated by Section 9 protocol during the three week test period. Additional samples were taken, more frequently than what the methodology of the test mandates.

Lead discharge was found to be low for the non-lead components. The overall mean lead concentration (adjusted to one liter) during the entire test period was below 5 µg/L for all non-lead components. For some of the conventional leaded components the mean was significantly higher and even exceeded the LCR Action Level. The corresponding "Q" statistic value was also higher than the LCR level for some of the leaded parts. In addition, lead discharge decreased by more than 50 percent over the three weeks of testing for the non-lead parts in the test water. The equivalent leaded components continued to leach about 70 percent of their initial lead discharge at the end of the test, indicating that the rate of decrease in lead dissolution was more significant for unleaded parts. No copper, selenium or bismuth leaching data were reported in that study. (Maas and Patch 2004)

Previous work carried out by the Environmental Quality Institute had quantified lead discharge of non-lead water service products using Envirobrass® II brass alloy. Non-lead ball valve curb stops and elbow couplings were examined, following a modified Section 9 protocol. The extraction water and pressurized manifold system were different than Section 9 procedures. The water samples were analyzed for total lead, but bismuth and selenium levels were also enumerated. Lead discharge for the parts tested was quite low. The one liter adjusted overall mean, for the whole testing period, was below 2.0 μg/L for all Envirobrass® parts. Selenium discharge was reported as barely detectable, even at week one of the test, whereas bismuth discharge was also very low. (Maas 2002)

Recent "dump and fill" tests conducted by the Greater Cincinnati Water Works (GCWW) examined lead leaching from non-leaded brass meters and fittings versus their leaded counterparts (De Marco 2005). In the case of brass meters after six hours of stagnation, the Envirobrass® leached 28 μg/L lead, which would correspond to a concentration of 4.9 μg/L if adjusted to one liter. The conventional brass meters leached as much as 1,240 μg/L, which translated to 94.7 μg/L if the contents were diluted into one liter. In tests of fittings containing between 0.2 and 8 percent leaded brass alloys and having the same water holding time of six hours, the non-lead materials showed significantly decreased lead dissolution into drinking water when compared to the leaded brass. Nonetheless in the case of curb stops, the lead released (adjusted to one liter) was above the LCR Action Level of 15 μg/L even for the 0.2 percent lead alloy (De Marco 2005). No copper, bismuth or selenium leaching data were presented.

A study by Sahoo et al. (1999) examined the corrosion behavior of Envirobrasses® II and III using weight loss and potentiodynamic polarization techniques. Experiments were undertaken in distilled water and in solutions with chloride and ammonium at pH of 6 and 8 and in wet clay. Based on the results, Envirobrass® II displayed corrosion characteristics similar to those of C83600 under all experimental conditions. Envirobrass® III and its leaded counterpart demonstrated corrosion resistance superior to that of C83600 and Envirobrass® II. Although these findings were not directly related to the propensity of the tested alloys to leach metals into drinking water, they can be considered preliminary results that are indicative of their general corrosion behavior.

Corrosion Propensity

Copper alloys, such as brass, are also subject to exotic modes of corrosion failure including intergranular corrosion, stress corrosion cracking or season cracking (Pugh et al 1966; Thompson and Tracey 1949) and dezincification (Lucey 1965). The likelihood of a brass device failing from any of these phenomena is a complex function of:

1. The alloy, including trace constituents such as arsenic at levels of 0.02 percent
2. How it is produced and in residual stresses within the metal
3. How it is installed, including mechanical stress on the device
4. Water chemistry.

There are standard test apparatus to investigate dezincification and stress corrosion cracking susceptibility of particular alloys (Lucey 1965; ASTM 1994). The water industry documented failures from chloramine attack on brass that were clearly inter-granular more than 50 years ago (Larson et al. 1956; Ingleson et al, 1949; anonymous 1951). It is noted that modern brass alloys in drinking water are also subject to this type of attack and failure. Depending on the aggressivity of the water and the alloy, the metal can essentially disintegrate with individual grains falling into the water from the device (Sundberg 2006; Sundberg et al. 2003). Ammonia is especially prone to causing stress corrosion cracking and intergranular attack (e.g. Guo et al 2002), but other nitrogen species and other constituents are clearly involved (Giordano et al. 1997; Rao and Nair 1998) These phenomena become more likely above about pH 7.0 and in one study they were maximized at pH 11.4.

For the case of non-leaded brass alloys, the same concerns stand regarding potential for exotic failure modes. Research on the dezincification of bismuth containing alternative brasses (1-3 % Bi) showed that these alloys were more susceptible to dezincification than leaded brass, when evaluated in synthetic tap water using potentiostatic polarization, scanning electron microscopy (SEM) and electron probe microanalysis (Kim et al. 2001). In other experiments, Envirobrass® I and II showed similar corrosion characteristics to leaded alloy C83600 when using potentiodynamic polarization and weight loss measurements (Sahoo et al. 1999). Both Envirobrass® I and II (5 % Zn) were not prone to dezincification, as evaluated by procedures such as "ISO 6509" and the "Hanningfield" test. Envirobrass® III (36 % Zn) had excellent corrosion resistance but was prone to dezincification (Sahoo et al. 1999).

With respect to stress corrosion cracking, one manufacturer states that cracking, either easily discernible (Figure 2.4a) or not visible to the naked eye (Figure 2.4b), has been observed during dry machining of copper-bismuth-selenium alloy surfaces of thin sections that were heated up. Heating can be caused due to excessive feeds and speeds or dull/ incorrect tooling. A conventional leaded alloy (C83600) exposed to the same machining process did not form cracks. However, the manufacturer reports that the same non-lead products could have been machined differently and not result in cracks. (Ford Meter Box Company 2005a).

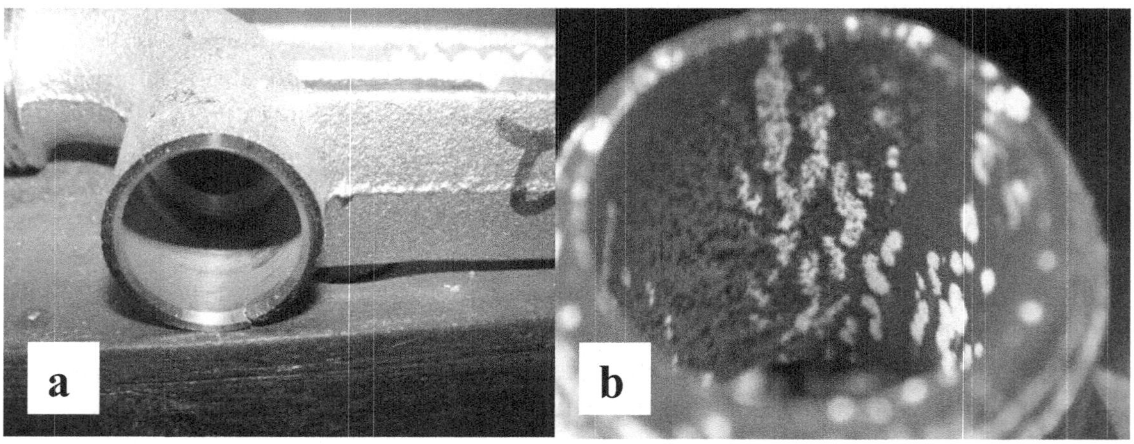

Figure 2.4. a) Envirobrass® II (C89520) casting after machining of solder port, with presence of a visible crack in the bottom of the machined surface. b) Envirobrass® II (C89520) casting, which did not show visible signs of cracks to the naked eye. Cracks were revealed as light green areas during a fluorescence penetration test. (Ford Meter Box Company 2005a)

In a lead-leaching experiment conducted at Virginia Polytechnic Institute and State University as part of this AwwaRF project, visible cracks in some non-leaded alloys were also observed after machining. During setup of the experiment, the alloys provided were cut with a lathe in order to obtain desirable sample shape and dimensions, based on the experimental design. This process caused cracks to form on a few unleaded brass specimens (Figure 2.5). The cracks were observed in a limited number of coupons and therefore do not constitute a generalized case. This may also be an isolated instance for the specific heat of material that was provided, or may have resulted from improper lathe use. Nonetheless, it does reinforce cause for concern. Results of this evaluation are contained in Appendix B of this report.

Figure 2.5. Unleaded brass coupon in the Virginia Tech leaching study. Cracks were observed after cutting of the original material to smaller coupons.

Other manufacturers assert that some alternative brass alloys possess excellent stress corrosion cracking resistance. More specifically, it is reported that "*the susceptibility of ECO BRASS® to stress corrosion cracking is one-thirtieth and below the usual materials according to the JIS ammonia test and CES ammonia test and solves the stress corrosion cracking problem that they have*" (Sambo Copper Alloy Company 2005). The JIS method involves soaking in an atmosphere of NH_4OH aqueous solution (concentration not less than 11.8 percent) for two hours at room temperature and identifying cracks by visual observation. The same manufacturer claims that the dezincification corrosion resistance of this alloy is also excellent, when evaluated using the "ISO 6509" test (Sambo Copper Alloy Company 2005). These issues are deserving of additional study.

ASTM SPECIFICATIONS FOR NEW NON-LEADED BRASSES

ASTM sets voluntary technical standards for materials, products, systems and services. These standards can theoretically draw on the wisdom of more than 30,000 technical experts including producers, users, consumers, government and academia from over 100 countries (ASTM International 2005). The specifications set by these standards offer detailed description as to how production should be made or as to how a process should be performed (NDT 2005).

ASTM standard specifications can play an important role in guiding design, manufacture and trade. They are developed by a large group of interested parties and are therefore widely accepted. From the manufacturers' perspective, a standard of production results in a more uniform and consistent process or product (NDT 2005). In addition, the specifications provide a basis for comparison between products that helps companies compete in the global marketplace. From the consumers' standpoint, ASTM specifications allow them to use products that have been tested and satisfy the requirements of the applicable standard.

ASTM Standard Specifications

No specifications have been developed by ASTM that focus specifically on non-leaded brass alloys. Some of the new alternative alloys have already been listed in existing ASTM standards for copper alloy applications, whereas some others are currently under evaluation for ASTM listing.

From the Envirobrass® family of alternative brass alloys, Envirobrass® I (C89510) and Envirobrass® II (C89520) are already listed in ASTM B30 Standard for copper alloys in ingot form, as well as in ASTM B584 Standard for sand castings (Table 2.8). From the FederAlloy® series of non-leaded brass alloys, currently C89844 is approved for the same ASTM Standards (B30 and B584), as well as B763 (Table 2.8). EnviroBrass® III (C89540) is expected to be included in 2006/2007 standards for Permanent Mold Castings (Sadayappan 2005a). Other alloys from the Envirobrass® and the FederAlloy® series that are currently commercially available have not received ASTM accreditation so far. EcoBrass® data is being collected for C87851 (sand casting) and C87852 (permanent mold casting) and the alloy is expected to be included in B30, B584 and B806 Standards in 2006/2007 (Chase Brass & Copper Co. 2006).

Table 2.8
ASTM specifications for non-leaded brass alloys (CDA 2000).

Alloy	Form	ASTM Standard	Description
Envirobrass® I (C 89510)	Ingot	B30	Copper base alloys in ingot form
	Sand	B584	Copper alloy sand castings for general applications
Envirobrass® II (C89520)	Ingot	B30	Copper base alloys in ingot form
	Sand	B584	Copper alloy sand castings for general applications
FederAlloy® 844 Version (C89844)	Ingot	B30	Copper base alloys in ingot form
	Sand	B584	Copper alloy sand castings for general applications
		B763	Copper alloy sand castings for valve applications

ASTM Standard B30 sets the requirements for copper alloys in ingot form to be re-melted for the manufacturing of castings. ASTM Standard B584 sets the requirements (e.g. minimum mechanical properties and chemical composition) for copper alloy sand castings intended for general applications. ASTM Standard B763 sets the requirements for copper alloy sand castings intended for valve applications: materials and manufacture, chemical composition and mechanical properties of suitable alloys are specified by this standard.

The composition requirements for the three ASTM accredited non-leaded brass alloys match their composition as presented previously in this chapter under the heading "*Description of the Composition, Lead Content and Mechanical Properties of New Non-leaded Brasses*". The minimum mechanical properties mandated by the relative standards are reported (Table 2.9).

Table 2.9
Minimum mechanical requirements for ASTM accredited non-leaded brasses. No mechanical properties are mentioned in the B30 Standard as ingot is an intermediate product.

ASTM Standard	Alloy	Minimum Tensile Strength ksi (MPa)	Minimum Yield Strength ksi (MPa)	Minimum Elongation in 2 in. or 50 mm (%)
B584	C89510 (Envirobrass® I)	26 (184)	17 (120)	8
	C89520 (Envirobrass® II)	25 (176)	17 (120)	6
	C89844 (FederAlloy® Version 844)	28 (193)	13 (90)	15
B763	C89844 (FederAlloy® Version 844)	28 (193)	13 (90)	15

SUMMARY OF CURRENT STATE EXPERIENCE WITH NON-LEADED MATERIALS

Plumbing codes [including the International Plumbing Code (IPC) and the Uniform Plumbing Code (UPC)] have been adopted by most states. These codes cover water supply and distribution issues, among others. They do not specifically exclude the use of lead in drinking water system components. In the USEPA Simultaneous Compliance Workshop (2004b) it was proposed that plumbing codes be altered so that they provide true "lead-free" materials.

Action has been taken by different states in order to eliminate lead use in drinking water components (e.g. California) or to promote use of components with lower lead content (e.g. Massachusetts). Although there are potential limitations to each of these initiatives, they do provide a potential model to consider at the national level, and are discussed below.

California

California is widely considered to be one of the most active states in reducing lead in plumbing components through Proposition 65 (most commonly referred to as CA Prop 65). CA Prop 65 was adopted as the "Safe Drinking Water and Toxic Requirement Act of 1986" as a means to prohibit the discharge of toxic chemicals at levels that would cause significant health concerns into "any source of drinking water" (OEHHA 2005). Faucet water was also included in the interpretation of drinking water sources (Weil 2005). With respect to lead, CA Prop 65 set the "safe harbor exposure level" at 0.5 ug/day (OEHHA 2005). This lead level takes into account cumulative exposure from multiple sources. It is thought to be based on data (presented in Canfield 2003), according to which a 10 μg/L blood lead level results in a 10 point IQ deficit in children (Weil 2005).

CA proposition 65 is entirely imposed through private citizen litigations (Weil 2005). In general, relatively few lawsuits go to trial, and settlements are often agreed upon by all involved parties. This statute has been utilized extensively, and as of 1998 (Heumann 2005):

- 950 Prop 65 notice letters had been sent to 3,710 businesses
- 200 lawsuits had been litigated
- $25 million had been spent in enforcement payments
- $250 million had been spent in reformulations/ process changes/ compliance.

Table 2.10 lists several lead-related Proposition 65 lawsuits that have been litigated.

Table 2.10
Major Proposition 65 Lawsuits, related to lead (from Weil, reported by Heumann 2005)

Product	Lawsuit Title	Court	Case Number
Faucets	People v. American Standard	San Francisco Superior Court	CGC-92-948017
Water Coolers	Mangini v. Haws Drinking Faucet Co.	San Francisco Superior Court	CGC-93-952872
Water Meters	Natural Resources Defense Council and Environmental Law Foundation v. Badger Meter	San Francisco Superior Court	CGC-97-984562
Galvanized Pipe Products	Mateel Environmental Justice Foundation v. Edmund Gray Co.	San Francisco Superior Court	CGC-00-309176
In-Line Plumbing Devices	As You Sow v. Conbraco Industries	San Francisco Superior Court	CGC-01-400568
Submersible Pumps	People v. Aeromotor Pumps	Alameda County Superior Court	733686-7

The issue of lead from brass devices was initiated in 1992, when elevated lead concentrations were measured in first-draw water samples from brass faucets. The State filed the "People of the State of California versus American Standard et al." case in 1992 against twelve defendants. Five separate consent judgments were agreed upon in 1995 and 1996. All settlements establish the same lists of covered and excluded products. In most settlements, covered products must meet the lead testing requirements of NSF/ANSI Standard 61, Section 9 (1994 version) with the normal Q statistic of 11 μg. Residential kitchen faucets (all settlements), and covered products (in some settlements) are to be tested according to the same requirements, but with a more strict Q statistic of 5 μg. Compliance schedules that eventually reach 100 percent of covered products sold in California are provided in each settlement. Certain penalties and payments were also agreed upon. The use of subsequent versions of Standard 61 for testing is at the discretion of the settling defendants. In addition, all compliance documents remain confidential. (Heumann 2005)

CA Prop 65 is imposed through enforcement actions seeking injunctive relief and civil penalties that are initiated by private citizens, non-profit groups or government attorneys. In addition, government attorneys can join actions initiated by private citizens or groups. Since California is one of only a few states that have not adopted NSF/ANSI Standard 61, Prop 65 settlements are the only means of enforcement for toxic contamination from indirect additives to drinking water.

Plumbing components meeting the standard for CA Prop 65 have been utilized for other drinking water components such as check valves, drinking water fountains and water meters. It is believed that devices that meet this standard may have a competitive advantage in the marketplace, even outside of California (Weil 2005). Labels for plumbing devices in other states often list that the products are Prop 65 compliant.

Natural Resources Defense Council and Environmental Law Foundation v. Badger Meter was filed in 1997 against six defendants. Three defendants settled early and the others agreed to a consent judgment in 2000. The settlement covers residential water meters with spuds of ¾ inch or less. Lead testing of meters was to be done according to NSF/ANSI Standard 61, Section 9 (1999 version) with the California revised Q statistic of 5 μg. It also defined new alloys with lead content of 0.25 percent or less, and identified two (FederAlloy® I (836) and SeBiLOY® II) as meeting the definition of a non-lead meter. Certain penalties and payments were also agreed to. The use of subsequent versions of Standard 61 for testing is at the discretion of the settling defendants. In addition, all compliance documents remain confidential

Another settlement initiated the implementation of a non-lead program by the Los Angeles Department of Water and Power (LADWP). The LADWP and the California Attorney General (AG) joined an existing action (filed under seal by the Environmental Law Foundation) in 1999. A settlement was agreed to by the LADWP, the California AG and a defendant in 2000, and the LADWP and California AG exited the litigation. The terms of the agreement financially supported the project, which aims to substitute old leaded water meters, curb valves, and other service line components with non-leaded products. Replacements were begun in April 2001 and the program is expected to result in replacement of all installed water meters in a time period of approximately 15 years. The annual replacement goal was initially set at 40,000 meters. A more detailed description of the LADWP program is contained in this chapter under the heading entitled "*Utility Motivations and Experience with Non-lead Brass Components*".

The Proposition 65 settlements resulted in more rigorous lead reduction methods for certain products, and the development and deployment of alternative alloys. As non-lead meters and fittings have been installed in California, a population of these new materials has been created that can be tested in the field. In addition, due to distribution channel constraints, essentially all residential kitchen faucets sold or distributed by major manufacturers in the U.S. now meet the testing provisions of NSF/ANSI Standard 61, Section 9 with a restricted Q statistic of 5 ug. In total Proposition 65 has reduced lead in some products, and provided encouragement for further lead reduction in others (Heumann 2005).

Massachusetts

The Massachusetts state plumbing code sets a requirement with respect to the maximum allowable lead content in drinking water system components. This regulation focuses on the use of materials with lower lead content for drinking water applications, rather than the complete elimination of lead. More specifically, the Massachusetts State Plumbing Code (248 CMR) requires that pipes, valves, pipe fittings, aerators or faucets which are utilized in drinking water applications contain no more than 3 percent lead as of January 2, 1995 (Laprade 2005). Until very recently, the Massachusetts Plumbing Board accepted NSF/ANSI Standard 61 testing as proof of compliance with the 3 percent lead provision. After learning that NSF/ANSI Standard 61 compliance does not assure that products contain less than 3 percent lead, the Board no longer accepts NSF/ANSI Standard 61 data and strictly enforces the 3 percent lead provision (Viola 2005).

UTILITY MOTIVATIONS AND EXPERIENCE WITH NON-LEAD BRASS COMPONENTS

By installing non-leaded brass components, utilities can demonstrate "due diligence" or the prevention of foreseeable harm. In some cases, utilities have initiated lead source replacement programs in response to regulatory requirements such as lead service line replacement programs, or customer concerns about the health effects of lead exposure. In other cases, utilities have embraced current technological advances such as non-lead brass meters and have incorporated them into their contract specifications and maintenance programs (i.e. meter replacement programs). Some utilities have implemented rebate programs to motivate consumers to purchase non-lead faucets and other fittings. In California, money available from Proposition 65 and other legal settlements is used to support these programs.

For many utilities, the switch to non-leaded components may also be based on the potential for improvement in the utility's 90th percentile lead levels. For utilities located in warm climates, the meter is generally located outside the residence (Figure 2.6), while for cold climates, the meter is located either in the basement or in a meter pit (Figures 2.7 and 2.8). The mass of lead leached from a leaded meter and associated fittings may potentially have more of an impact on lead levels measured at the tap in cold weather settings. When compared to warm climate service configurations, cold climate service installations may exhibit a shorter distance to the faucet and therefore less dispersion of the lead mass.

In 2000, Asarco commissioned a survey of U.S. water utilities to identify utility plans for replacement of leaded components with non-leaded components. Asarco is a smelting and refining company that produces the bismuth selenide alloy used in making Envirobrass®, and was part of a consortium that developed bismuth selenium compounds as replacements for lead in brass. They received responses from 301 utilities nationwide, with 50 percent of survey respondents representing small systems (defined as serving <5000 connections). Survey findings show that 20 percent of utilities (61 of 301 responses) currently have plans to replace their lead-containing brass components. Twenty-one percent of survey respondents indicated they were not aware that non-leaded brass options are available (Asarco 2000).

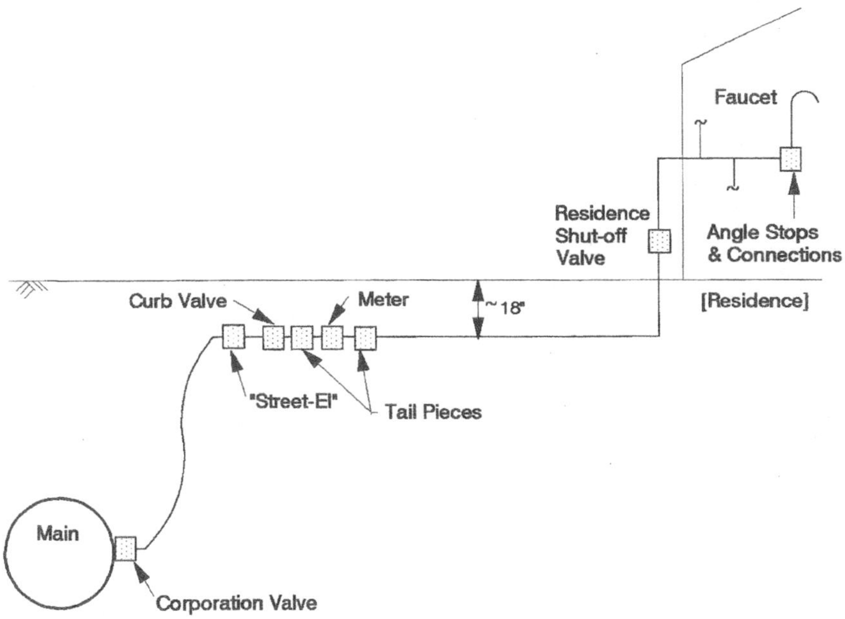

Figure 2.6. Diagram of Typical Warm Climate Residential Service (Provided by Heumann 2005)

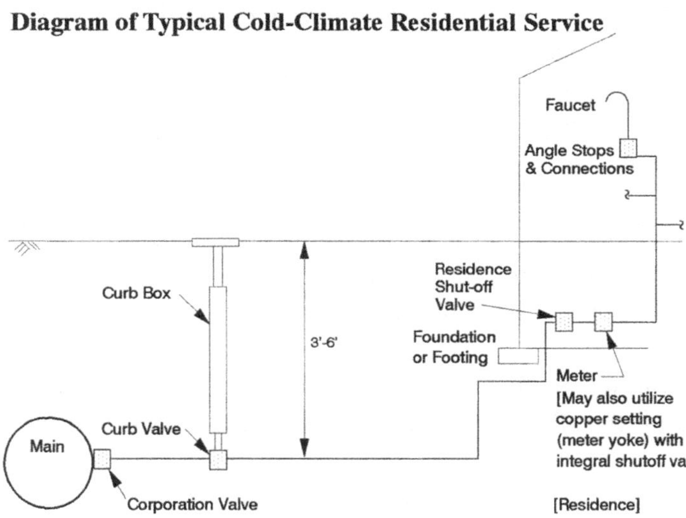

Figure 2.7. Diagram of Typical Cold-Climate Residential Service (Provided by Heumann 2005)

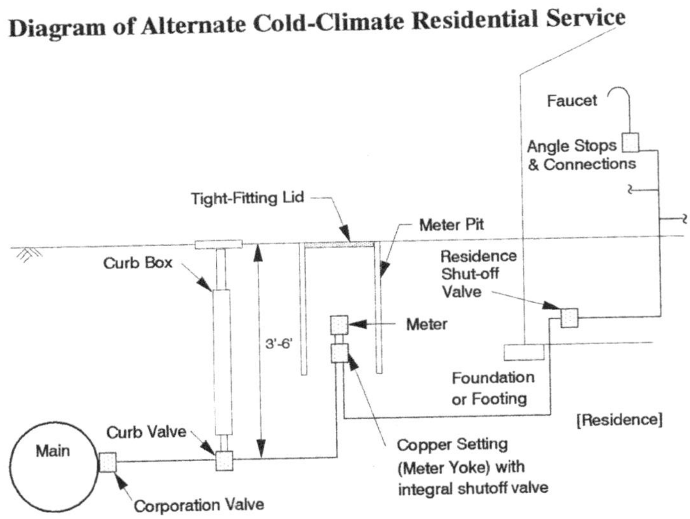

Figure 2.8. Diagram of Alternate Cold-Climate Residential Service (Provided by Heumann 2005)

A water utility survey conducted as part of AwwaRF Project 3018, *Contribution of Service Line and Plumbing Fittings to Lead and Copper Rule Compliance Issues*, provides some insight into utility lead component replacement programs (HDR/EES 2005a). The survey captured utility knowledge on lead sources and the availability of utility data and information related to project objectives. Preliminary survey results applicable to this project (based on 30 responses) are summarized as follows:

- Three respondents have programs to install non-lead meters and other fittings.
- Ten of 30 survey respondents have developed their own specifications for either lead-free (<8 %) or non-lead (<0.25 %) components.

Table 2.11 provides examples of utility lead source replacement programs that utilize non-lead brass components, and the following sections provide additional details on these programs.

Los Angeles Department of Water and Power

LADWP has long been active in efforts to reduce lead contamination in drinking water. Beginning in the early 1980s, studies were performed to determine corrosion rates on water system materials, including the development of weight-loss corrosion measurement methods and sampling protocols. In order to assess the effect on residential plumbing, an employee residence sampling program was conducted in 1985. This resulted in a change to lead-free solder for all LADWP activities in advance of federal regulations. Studies of lead-dispersion from lead goosenecks were also carried out, resulting in a program to identify and remove lead goosenecks (although not required by federal regulations). Removal of all known lead goosenecks from distribution system was completed in 2005. In addition, corrosion inhibitor effectiveness studies on LADWP water sources were performed, and a pilot scale corrosion control facility was constructed in 1990, which continues in operation. A demonstration scale corrosion control facility is pending. These activities have been pursued, even though all LCR residence compliance samples collected have been below the federal action levels for lead and copper.

Table 2.11
Utility Programs Utilizing Non-Leaded Brass

Utility	Program Description (Year Started)	Reference
Los Angeles Dept. of Water and Power	Replacement of substandard curb valves and meters as a result of legal settlement with supplier.	Maas, Patch and Lagasse 2002; Heumann 2005; HDR/EES 2005b
East Bay Municipal Utility District	Old meter and service line fittings replaced with non-leaded parts by attrition.	HDR/EES 2005c
Bangor Water District, Maine	Large meter replacement program; utility specified non-leaded brass components on new contracts.	Moriarty 2005
Greater Cincinnati Water Works	Meter replacement program	USEPA 2005b
Seattle Public Schools	Replacement of drinking water fountains, faucets, flexible connectors and shut-off valves.	Seattle Public Schools 2006 Pierson, 2006 Boyd et al 2005
San Francisco Public Utility Commission	Faucet rebate program; meter replacement program, specification of non-leaded curb stops.	HDR/EES 2005d

The LADWP initiated a "Non-lead Brass" Program as a result of a legal settlement regarding substandard valves and other parts supplied to the utility. These components contained up to 40 percent more lead than allowed by contract specifications. In 1998 LADWP tested 195 parts supplied under contract and found that 68 percent failed to meet contract specifications. The measured lead levels in drinking water did not result in water quality problems to LADWP's customers or violate any USEPA action levels for lead. Settlement funds were used to initiate the program.

LADWP uses two different alloys in the "Non-lead Brass" Program. The alloy used for waterworks fittings (curb valves, meter tailpieces, etc.) is Federalloy® I (version 836) with a 0.1 percent maximum lead content. LADWP testing of production pieces revealed a range of lead concentrations from 0.07 percent to 0.09 percent. The alloy used in meters is SeBiLOY® II (a.k.a. Envirobrass®), with a 0.25 percent maximum lead content. LADWP testing of production pieces revealed a range of lead concentrations from 0.12 percent to 0.18 percent. These testing results indicate that both FederAlloy® and SeBiLOY® could probably meet a 0.2 percent lead limit. LADWP testing results indicate that the percent elongation for FederAlloy® under failure testing more closely approximates the behavior of traditional brass than does SeBiLOY®. However, there have been no observed differences as regards failures of either non-lead alloy when compared to traditional brass over the past 4.5 years. This is true despite the fact that soils in the LA area are active, providing a relatively rigorous application environment.

LADWP began installing only non-lead meters in 2001 and intends to replace all meters (~700,000) by 2016, with an annual replacement goal of 40,000 meters. As of July 5, 2006, there were 723,250 service connections (88 percent of which are residential) and more than 171,000 meters have been replaced with non-leaded meters. Meter replacement normally includes replacement of meter couplings (a.k.a. tail pieces Figure 2.6), so approximately the same number of tail piece pairs has also been replaced. In addition, when a meter is replaced, the curb valve is sometimes replaced. Since 2001, more than 70,000 curb valves have been replaced. This program will continue for the foreseeable future (Heumann 2005; HDR/EES 2005b; Maas, Patch, and Lagasse 2002).

With regard to the new non-leaded meters that have been installed, the LADWP has not experienced failures or leaks at a higher frequency than that observed when old leaded meters were installed. At the same time however, failures early in the product life are expected to be rare. The LADWP has started to perform tests on the structural stability and leaching characteristics of the non-lead service line components that have been retrieved from the field after approximately 5 years of service. The retrieved meters have exhibited no evidence of structural weakening or dealloying. Testing for changes in leaching characteristics are in progress, and although that work is not yet complete, there have been no significant changes in leaching discovered. The results of testing will be published eventually (Heumann 2005).

East Bay MUD

Between 1985 and 1992, East Bay Municipal Utility District (EBMUD) in California proactively replaced known lead service lines. In 2002, EBMUD implemented a program to further reduce the potential sources of household lead through the replacement of brass meters and service line fittings. This is an attrition program that focuses on removing the oldest equipment from the system first. EBMUD replaces obsolete meters and fittings with California Proposition 65 compliant non-lead meters and fittings. Each year EBMUD issues contracts for the replacement of approximately 15,000 small water meters, 150 large water meters, and 15,000 curb stops. EBMUD uses non-lead materials specifications that require the installation of meters and fittings whose components, individually, do not exceed a lead content of 0.25 percent (HDR/EES 2005c). EBMUD's most recent 90th percentile level for lead was 5 μg/L.

Bangor Water District

In the year 2000, the Bangor Water District in Maine decided to start specifying non-leaded brass meters (<0.25 percent lead) as part of the planning process for replacing all large meters in the system (Moriarty 2005; USEPA 2004b). The District serves approximately 10,000 services using a surface water source that is treated by disinfection with ozone and chloramines, and pH adjustment with soda ash. Although the District's 90th percentile samples have a lead concentration below the action level (June 2004 90th percentile value was 10 μg/L), the District is motivated to reduce sources of lead that could potentially leach into the drinking water. The

District has no lead service lines but has identified meters as a potential source of lead. A limited meter testing program in year 2000 showed the value of using non-lead components. The utility's Materials Specifications (6-1-2004) include non-lead materials for corporation stops, curb stops and miscellaneous brass goods (Bangor Water District 2005).

Greater Cincinnati Water Works

The Greater Cincinnati Water Works (GCWW) recently started to specify non-leaded materials in their bid documents (USEPA 2005b). Although GCWW's 90th percentile tap samples have never exceeded the lead action level (most recent 90th percentile value was 7.3 μg/L), GCWW is committed to reducing sources of lead in distribution system materials. With respect to leaded brass meters and components, the utility's approach is to change them out as part of normal replacement cycles. For example, the utility is currently replacing meters in order to implement automatic reading so they have specified non-lead meters.

Seattle Public Schools

The Seattle Public Schools includes 97 buildings (schools and related facilities), all are served with water from Seattle Public Utilities. Seattle Public Schools conducted a comprehensive testing program in 2004 to identify existing endpoint devices (i.e. water fountains, classroom sinks, kitchen taps) with elevated lead levels. Initially, the testing program was focused on meeting the USEPA guideline of 20 μg/L for lead in school drinking water from first draw and flushed samples, but the school system later established its own lead action level of 10 μg/L in December 2004. All non-compliant sources have been or will be either fixed or taken out of service. Prior to mitigation, extensive testing at more than 3,100 drinking water fountains and sinks showed that about 20 percent exceeded the USEPA guideline (Seattle Public Schools, 2006). Seattle Public Schools is currently completing remediation of endpoint components that exceeded the 10 μg/L action level. Remediation plans developed for each individual school include full or partial pipe replacement, replacement of drinking water fittings (i.e. water fountains, faucets), replacement of flexible connectors and shut-off valves, and in some cases installation of point-of-use filters that are NSF certified for lead removal. Where brass components were replaced, non-leaded ($\leq$0.03 percent lead) brass components were utilized. As of March 2006, 87 of 97 school buildings had completed remediation and confirmation sampling was performed to demonstrate that lead levels are less than the SPS-specified action level of 10 μg/L. These schools have resumed use of City water and have removed temporary supplies of bottled water. Remediation work is continuing at the remaining 10 schools and is anticipated to be complete by the end of 2007 (Pierson 2006; Boyd et al 2005).

San Francisco Public Utility Commission (SFPUC)

In 2001 and 2002, the SFPUC's Water Quality Bureau conducted bench-top lead leaching tests to better understand the sources of lead (consumer's plumbing vs. water system's plumbing) and to assess the merits of switching to non-leaded in-line components. The testing program in 2001 focused on leaded and unleaded water meters (5/8-inch, 1-inch and 2-inch) and curb stops (1-inch to 2-inch). The 2002 program evaluated four types of unleaded faucets. Bench testing results indicate that (i) there are significant differences in the amount of lead leached between brands of unleaded meters; (ii) curb stops have similar or higher lead leach rates as meters; and (iii) some unleaded faucets leach more lead than the others.

In early 2003, the SFPUC initiated a pilot-scale program to test non-leaded curb stops under field conditions to evaluate operations and maintenance issues. As part of the pilot program, twenty 1-inch lead-free curb stops were installed in new residential installations in the City. After 1-year of installation, these curb stops were reportedly functioning well.

In 1998, SFPUC's Water Quality Bureau submitted a California Public Health Goals Report to the SFPUC proposing several strategies for reducing lead exposure from drinking water. One of the proposed strategies was to implement a faucet replacement program. In 1999, a pilot faucet replacement program was implemented to reduce lead exposure from drinking water to young children at City preschools and daycare centers. In 2000, the SFPUC expressed an interest in continuing the giveaway program to daycare centers, high lead sites, schools, and recreation centers. Due to the success of the pilot faucet replacement program, SFPUC initiated a faucet distribution program for residential customers in 2001. Customers were given the opportunity to purchase a specific kitchen faucet chosen by the SFPUC for \$10, a savings of \$180. The program objective was to enhance the general public's awareness of the potential lead leaching problem from residential plumbing fittings. Since 2001, 3140 non-leaded faucets have been distributed to residential customers and 487 faucets have been given free to preschools/daycare centers. In 2002, SFPUC gave 505 non-leaded faucets and drinking fountains to 47 City schools.

In December 2000, SFPUC adopted a resolution to modify the existing water meter replacement program by discontinuing the purchase of leaded brass water meters whenever a suitable unleaded substitute is available. In 2002, the SFPUC implemented a non-leaded meter replacement program with a goal of replacing 7,000 to 8,000 meters annually over a 20-year period. As of October 2005, about 13,000 non-leaded meters have been installed (HDR/EES 2005d).

POTENTIAL IMPACTS TO MANUFACTURERS

A number of new cast alloys were developed with little or no lead content specifically to meet NSF requirements. An industry consortium of the CDA, AFS, BBIM, and Materials Technology Laboratory of CANMET in Canada conducted research and developed a series of copper alloys which contain 0.1 to 0.25 percent lead (commercially available as EnviroBrass® also known as SeBiLOY®). The Federal Metal Company has concurrently developed an alloy containing 0.1 percent lead (commercially available as FederAlloy®) and the Sambo Copper Alloy Co., Ltd developed a silicon brass with the trade name ECO BRASS®. No lead is added to these alloys, and the minimal amount of lead present is due to the lead contained in recycled materials and ore that are used as the source metals. These alloys are formed by substituting selenium and/or bismuth for lead in the case of EnviroBrass® and FederAlloy®, and silicon and phosphorus in the case of ECO BRASS®.

There is currently no problem with supply or long-term availability with respect to materials used in the manufacture of ECO BRASS®, however there are impacts to manufacturers who utilize bismuth and/or selenium alloys to create a variety of drinking water components as described below.

Supply and Demand for Bismuth and Selenium

Some studies are showing that bismuth and selenium could be in short supply with competing European and Japanese demands (USEPA 2005b). The result of a short supply would be price increases, as demonstrated by the recent increase in the spot price for imported selenium as illustrated in Table 2.12 below.

Table 2.12
United States Selenium Supply and Demand Statistics

	2000	2001	2002	2003	2004
Imports for Consumption (metric tons)	4760	4830	4220	3670	3900
Exports, metal, waste, scrap (metric tons)	820	410	850	2430	900
Price ($/lb)	$3.84	$3.80	$4.27	$5.68	$27.00

Source: Summarized by Weed (2005) from data sources: George 2003a; USGS 2005a.

World production and consumption data for selenium is limited. Selenium is a rare element, widely distributed within the Earth's crust, but not in concentrations high enough to justify mining solely for its content. It is recovered as byproducts, mostly from the anode slimes associated with electrolytic refining of copper. World refinery production of selenium is estimated to have remained stable. The average global consumption of selenium during the past 4 years is estimated to have been about 2 million kilograms per year, 4 – 5 times the amount consumed in the U.S. More than one-half of the metallurgical selenium is used as an additive to cast iron, copper, lead, and steel alloys. The driving force behind the worldwide demand increases for selenium is the growing economy of China. In mid-2003, China started importing large quantities of selenium, which caused the price to soar as illustrated in Table 2.12. Currently, there is little to no selenium being sold because there is no supply. Additional statistics on selenium supply and demand are presented in the following table.

Bismuth consumption in the United States has ranged between 1930 to 2410 metric tons (4.5 and 5.5 million pounds) for the last five years and historically, prices have been relatively stable (United States Geological Survey (USGS) 2005a, b). USGS believes that world demand for bismuth is increasing approximately 5 percent per year. International agreements to eliminate lead from solder in manufacturing processes by 2005 in Europe, Japan, and North America will tend to increase the demand for bismuth over the next several years." (Weed 2005)

Bismuth is not produced domestically, so the U.S. relies on imported material to satisfy demand. Only a small amount of bismuth is obtained by recycling old scrap. Over the past 5 years, the U.S. has imported 40 percent - 50 percent of the world's refining capacity for bismuth. Throughout most of the world, bismuth is a byproduct of processing lead ores; in China, it is a byproduct of tungsten ore processing. The Tasna Mine in Bolivia, the only mine that produced bismuth from a bismuth ore, has been on standby since the mid-1990s awaiting a significant rise in the metal price." (USGS 2005b) Additional statistics on bismuth supply and demand for the period 2000 to 2004 are presented in Table 2.13.

Table 2.13
United States Bismuth Supply and Demand Statistics

	2000	2001	2002	2003	2004
Imports (metric tons)	2,410	2,220	1,930	2,320	2,100
Consumption (metric tons)	2,130	2,200	2,320	2,120	2,420
Price ($/lb)	$3.70	$3.74	$3.14	$2.87	$3.10
Imports as percent of Consumption	95%	95%	95%	95%	90%

Source: Summarized by Weed (2005) from data sources George (2003b); USGS 2005a and 2005b.

Processing Issues

Developers of the new non-leaded alloys state that the foundry characteristics (melting and casting) and their mechanical properties are similar to existing leaded alloys; however the new alloys may exhibit lower ductility (percent elongation) and lower impact strength. According to the CDA, all of these alloys have worked well in the appropriate applications (Weed 2005). Ford Meter Box (2005a) offers the perspective that the mechanical properties and manufacturability of the new alloys are somewhat inferior to the copper-lead alloys. Ford Meter Box is concerned about applications where the product is buried underground and failure would be difficult to detect.

The CDA (Weed 2005) states that there is no problem producing bismuth-copper alloys in foundries. However, some process changes may be required. If an operation produces both leaded and bismuth brass alloys, it is possible to produce both in the same furnace, however the wash heats necessary to condition the furnace for "non-lead" can be expensive and difficult to control. This often leads to dedicating specific melting equipment just for non-lead alloys and may even require the purchase of new equipment. Brass scrap that contains bismuth must be isolated from other brass scrap. Bismuth makes hot rolling and hot forming nearly impossible in copper alloys so bismuth scrap must be either re-used or sent to a foundry casting bismuth alloys. All brass mills in the U.S. test scrap for bismuth content and will refuse any load that contains bismuth. Presently, bismuth containing brass scrap is well controlled and used as run-around scrap or returned to a foundry. There may be a concern in the future as more consumer scrap enters the recycling stream and additional time and expense are required to identify and segregate bismuth alloys from other copper alloys.

Component Manufacturing Issues

The need to comply with NSF/ANSI Standard 61 is causing manufacturers to consider alternative strategies for production of brass water meters and other components in drinking water systems. The primary compliance strategies used to meet NSF/ANSI Standard 61 are to reduce the lead content of the brass alloy; reduce the surface area available for water contact; use a different material for water contact such as platings or coatings; or remove lead from the surface of the water contact component. (Weed 2005). The following section discusses some potential impacts of using non-leaded alloys to manufacture drinking water components.

The machinability of various alloys is compared to a standard machinability rating of 100 set for alloy C36000 (free cutting brass). Machinability ratings range from 75 to 85 for EnviroBrass® and FederAlloy®, and decrease as the bismuth content of the alloy decreases. The Copper Development Association states that EnviroBrass® III can be machined using essentially the same practice as leaded yellow brass, but there might be higher tool wear with EnviroBrass® I and EnviroBrass® II when compared with C83600 or C84400 alloys, and machining fluids may be needed (CDA 2005; Sadayappan 2005a). ECO BRASS® has a machinability rating of about 70 - 80 and other silicon brasses and bronzes have somewhat lower machinability ratings, 30 – 50.

There may also be issues associated with larger scale production. From a manufacturing point of view, if bismuth is used as a replacement for lead in alloys, this change involves upfront costs, design changes, and practice modifications. This could have large capital implications for manufacturers. The machinability ratings are based upon the speeds (rpm of the brass material) and feeds (inches per second feed rate of the tools) compared to alloy C36000, which has a rating of 100. So while another alloy may fall into the "free machining" class with a rating of 75, that means that the speeds/feeds rate at which material can be removed is only 75 percent of that of alloy C36000. The labor and overhead cost of making a part may be higher for an alloy rated at 75, compared with the C36000. Also, the ratings relate to the ability to remove the material, but do not always reflect the ability to produce the smooth surface finishes. Therefore, before significant investment is made, manufacturers need to know USEPA's direction on possible health impacts of bismuth (USEPA 2005b).

In yellow brass, used for faucets and other fittings, the addition of bismuth can cause cracking issues. Like water, bismuth migrates into cracks and expands by 3 percent as it solidifies. There is concern about the impact in field use or machining. Ford Meter Box (2005a) raises a concern with the generally lower ductility of the no-lead alloys and their demonstrated propensity to crack. The "...low ductility gives rise to concerns of failure, especially in thin sections of products that may be stressed in installation or usage. Indeed, manufacturers have had to redesign products to remove thin sections in order to reduce concerns with the low ductility. The propensity to crack has been found in the copper-bismuth-selenium alloys. The cracking has normally been found to occur on dry machined surfaces of thin sections that were heated up due to any of several reasons including excessive feeds and speeds or dull or incorrect tooling. The cracking, in some instances, is easily seen and discernible." (Ford Meter Box 2005b). Ford Meter Box has decided to not use these alloys as a solution to meet lower lead levels and instead is utilizing a lead-removing wash system known as the Kunkel Process to meet the NSF/ANSI 61 standard (Ford Meter Box 2005c). The Kunkel Process is a patented process that exposes parts to a series of acid/caustic solutions in order to dissolve lead contained on the surface of the part.

The decision to utilize non-leaded brass alloys to manufacture drinking water components involves a variety of criteria, not just the ability to meet lower lead levels. Ford Meter Box (2005b) provides perspective on the issues related to the manufacture of drinking water components, and the use of non-leaded alloys: "The selection of alloys must be based on product line configuration, manufacturing capabilities, resources and customer needs as well as safety. Product performance characteristics such as durability, machinability, resistance to corrosion, functionality, and strength are just a few items that must be considered when selecting a material for use in a product." Selover (2005) states that companies that manufacture components use several product performance criteria when selecting materials and developing their product design, including:

- Burst Strength/Water Hammer Resistance
- Bending/Torque Strength (mounting issues)
- Durability –500,000 on-off cycles
- Machinability –tolerances necessary to provide smooth, durable function
- Functionality in High Mineral Content Water
- Ability to Apply Decorative Finishes
- Abuse/Vandalism –Excessive Force
- Hot/Cold Water Temperatures

Presumably, non-leaded alloys used for manufacturing components would be subjected to these criteria as well.

Health and Environmental Impacts

Manufacturers may be gaining a competitive advantage by meeting the California Proposition 65 standard (USEPA 2005b). As a result of this standard, there have been major changes in the lead content of sandcast brass. Side benefits to reducing lead content include reduced worker exposure, lower air emissions, and less production of hazardous waste. However, Ford Meter Box (2005b) points out that waste streams containing selenium may be considered hazardous.

When the non-leaded brasses "…are machined dry (without cutting fluids), no significant lead-bearing particles enter the machine shop air. This means no special air purification equipment is needed, enabling a possible drop in insurance costs for manufacturers of plumbing fittings because their workers may not be exposed to lead. In addition, for sand casters, the costs for prevention of airborne lead and for reclamation and disposal of foundry sand should be lower." (CDA 1996).

IMPACTS TO PLUMBING INDUSTRY

The impacts and implications of moving the plumbing industry towards use of non-leaded brass components for drinking water involves a complex interaction among the groups involved: manufacturers of components, organizations that certify components and develop standards and codes for their use, inspectors that enforce those codes, and the installers of components (both utilities and private plumbers). There are several issues that can be considered including; development of non-leaded components, code revisions and code enforcement (at the utility, local, and state level), coordination between federal, state and local regulatory agencies and code organizations, education and training of plumbing inspectors and component installers, and standardization of terminology between groups.

Plumbing Manufacturers

Issues related to the development and manufacture of non-leaded components are contained in the previous section on manufacturing. In summary, they include:

- Increased manufacturing costs due to re-tooling, product redesign, reduced manufacturing efficiency and scale up of facilities to produce non-leaded components
- Concerns about cracking due to lower ductility of some non-leaded alloys, and the need to significantly re-design products
- Ability of non-leaded components to meet consumer, regulatory and/or companies' product performance criteria.

Possible changes to drinking water regulations, material certification requirements, and standards and codes to incorporate non-leaded brass alloys could all play a significant role in product development and the timing for implementing changes.

Standards and Plumbing Codes

In addition to the requirements found in the NSF/ANSI 61 standard, the occurrence of lead in materials that come in contact with drinking water is controlled by building codes such as the Uniform Plumbing Code, standards by the American Water Works Association (AWWA), the American Society of Mechanical Engineers (ASME), and ASTM. Plumbing products within buildings are generally regulated at the state or local level through plumbing codes. These local codes are normally based on Model Plumbing Codes such as UPC, IPC, and the National Standard Plumbing Code (NSPC). All plumbing codes require products to comply with an applicable product standard that contains minimum performance requirements and test procedures that impact product design, manufacturing process and cost (Viola 2002). The UPC, IPC and NSPC all reference NSF/ANSI Standard 61 for pipe, fittings, and faucets but do not reference NSF/ANSI Standard 61 for in-line valves. The ASME A112.18.1 standard for faucets and other end-point devices in the U.S. and CSA B125.1 standard for faucet and other end-point devices in Canada require end-point devices that convey drinking water to be in compliance with NSF/ANSI Standard 61. These standards are referenced in virtually all model and state- and locally-written codes. Most faucets sold at the retail and wholesale levels in the U.S. are certified to NSF/ANSI Standard 61 (USEPA 2005b).

Moving the plumbing industry towards the use of non-leaded alloys and components will likely require that certification, standard specifications, and code revisions be implemented. Minimum performance standards and certification evaluation procedures would need to be developed using realistic time-frames and schedules for implementing changes. The same would be true for plumbing code changes at the national, state, and local level.

There is often little or no discussion between the State drinking water primacy agency and the people working on plumbing codes (USEPA 2005b). It would be helpful to strengthen this connection, but this can be difficult, particularly in some states that have many jurisdictions. There needs to be increased dialogue between those that develop plumbing codes and those that implement them.

Installation and Inspection

Installers of non-leaded brass components (utilities as well as master and journeyman plumbers) will need to be educated about the availability of these components and any technical information needed for proper installation. There are several unknowns related to installation of non-leaded components, such as what the impact of installation and workmanship may have on drinking water quality, and how non-leaded components perform in the long-term when compared to historically used components that contain lead. Plumbing code inspectors may also need to be educated about revisions to plumbing codes in order to enforce code requirements for non-leaded brass components.

There is also a terminology gap between the various groups. In plumbing industry terminology, a fixture is a receptor that receives potable water after discharge (USEPA 2005b). By definition, a fixture does not contain potable drinking water, and therefore is not subject to lead in drinking water regulations. Current drinking water regulations may need to be revised to reflect industry terminology and avoid further confusion.

IMPACTS TO UTILITY OPERATIONS AND MAINTENANCE PROCEDURES

Utilities participating in two AwwaRF projects (Contribution of Service Line and Plumbing Fixtures to Lead and Copper Rule Compliance Issues and Performance and Metal Release of Non-leaded Brass Meters, Components and Fittings) were contacted regarding their experiences with installing non-leaded brass meters and components and any specific impacts on O&M issues. Because the installation of non-leaded brass components has only taken place within the last 5 years in the water industry, there is a lack of detailed information available. Utility experiences with material performance and installation costs are detailed in the following sections.

Material Performance

LADWP staff prefers FederAlloy® because the product material is more ductile, approximating that of the leaded brass that it replaced (Heumann 2005). FederAlloy® now has ASTM approval for two alloys: FederAlloy® I (836), which has been used by LADWP for fittings since 2001, and FederAlloy® I (844 MOD), which is designed for use in water meters. LADWP was also recently informed that the SeBiLOY® II, used in water meters purchased by LADWP since 2001, will be replaced by BiAlloy for water meters supplied under existing contract. This change was agreed to by NRDC as fulfilling the California Proposition 65 requirements, and the provisions of the 2000 Consent Judgment that helped initiate the LADWP meter replacement program. LADWP testing of the new materials has not yet been completed. It is expected that competition among all non-lead meter materials will ensue in the near future.

Since LADWP began replacing residential water meters in 2001, the utility has not seen fracture failures or leaks at a frequency different from older brass meters. Failures are not expected in the early years of the program. Meters retrieved and tested after 4.5 years of service had no structural weakening. In early 2003, SFPUC initiated a pilot program to test unleaded curb stops under field conditions to evaluate operations and maintenance issues (HDR/EES 2005d). As part of the pilot program, twenty 1-inch lead-free curb stops were installed in new residential service connections. After one year of installation, these lead-free curb stops were reportedly functioning well. However, the testing will go on for another year before a decision is made on their serviceability.

In 2005, the SPS replaced more than 360 drinking fountains from 97 separate schools and administrative buildings as a result of a comprehensive testing program in 2004 that identified endpoints with elevated lead levels (Seattle Public Schools 2006). Initially, the testing program was focused on meeting the USEPA guideline of 20 μg/L for lead in school drinking water from first draw and flushed samples, but the school system later established its own lead action level of 10 μg/L in December 2004. The replacement bubbler heads are "non-lead" brass construction with less than 0.2 percent lead; however, the presence of surface residuals with lead contaminated dust and cutting oils is common (ASTM 2005). Flushing is typically used to remove the surface residuals. For Seattle schools, a bubbler head pre-conditioning rig was fabricated and installed at an SPS operations facility. All bubbler heads are pre-conditioned before installation in schools. The pre-conditioning involves a 1-minute flush at 2 gpm (to remove any small metal particles from the casting/machining process), followed by a cumulative flush for 24 hours at 0.5 gpm. Upon installation, the bubbler heads are flushed for 5 minutes on-site before placing the bubbler head into service.

Installation Costs

Several utilities have estimated the costs of using non-leaded meters and other components as compared to components made using leaded brass materials. These cost increases range from 2 percent to 100 percent as illustrated by the following case examples for EBMUD, LADWP, GCWW and SFPUC.

EBMUD estimates that the specification of non-leaded meter replacement materials resulted in a 7 percent increase ($59,150) in program cost based on fiscal year 2005 figures (HDR/EES 2005c).

In 2001, LADWP compared meter replacement costs for both leaded and non-lead brass including the meter, curb valve and tailpieces as summarized in Table 2.14. The non-lead brass meter replacement costs were 30 percent higher than the leaded brass alternative, with installation done by the utility (Heumann 2005).

Table 2.14
Cost Comparison for High and No Lead Bronze Meters Installation at a Typical Service at LADWP (materials only)

Brass component (1 inch)	High lead bronze (85-5-5-5 bronze)	No lead bronze	Cost difference
Curb valve	$17.46	$19.17*	$1.71
Tail piece 1	$3.86	$6.64*	$2.78
Tail piece 2	$3.86	$6.64*	$2.78
Water meter	$18.04	$23.80†	$5.76
Totals	$43.22	$56.25	$13.03 (30%)

* FederAlloy® † SeBiLOY®

The following table presents the average full costs (i.e. with overheads) for new 1-inch service line installations at LADWP, based upon over 800 installations from July 1, 2005 to March 31, 2006. Labor hours and tool expenses are the same for non-lead and leaded brass installations done in the past. Materials are broken down into categories for "non-lead materials" and "other materials". LADWP is only allowed to charge customers for costs incurred, and the total service line cost corresponds to the current charge for a new 1-inch service line. The service line cost is also calculated for leaded brass, using a non-lead materials cost differential of 30 percent (as determined by LADWP on comparative bids in 2001). The service line cost increase attributable to the change to non-lead brass is 2 percent (see Table 2.15 and notes).

Table 2:15
Average Full Costs for Service Installations at LADWP

Cost Element	Total Cost† Conventional Materials	Total Cost† Non-leaded Materials
Labor	$1,190	$1,190
Materials (293)	254	293
Equipment	195	195
Construction Support	251	251
Miscellaneous‡	52	52
Total Service Line Cost	$1,942	$1,981
Relative service cost line w/ non-lead:		102%

* Data from 827 new service line installations
† Direct costs with overheads applied
‡ Other cost elements, not disaggregated

GCWW compared installation costs for leaded brass and 0.25 percent lead (non-lead) meters and fittings, and found that the materials cost would be 20 percent higher for the non-lead fittings installed by the utility and 27 percent higher for components typically installed by the homeowners (Table 2.16) (USEPA 2005b).

Table 2.16
Material Cost Comparison for Leaded and Non- Leaded Meter and Fittings at GCWW

Brass component (3/4 inch)	8% Lead	0.25% Lead	Cost difference
	Costs to Utilities:		
Ferrule	$20.60	$24.72	$4.12
Curb stop	$33.09	$39.70	$6.61
Ins coupling	$22.00	$26.40	$4.40
Total Cost	$75.69	$90.82	$15.13 (20%)
	Costs to Homeowners:		
DA Valve	$14.98	$28.08	$13.10
PWA Valve	$13.60	$25.80	$12.20
Expander	$7.96	$14.93	$7.02
Meter	$71.00	$68.00	-$3.00
Total Cost	$107.54	$136.81	$29.27 (27%)

SFPUC has provided cost information on non-leaded curb stops and faucets. In 2003, SFPUC initiated a pilot-scale program to evaluate non-leaded curb stops in residential service connections. Based on installation of twenty 1-inch non-leaded curb stops, the cost was reportedly twice the cost of existing curb stops used by the SFPUC (HDR/EES 2005d).

Summary

There is little documented information on the material performance of non-leaded components that have been installed in the field, however limited information indicates no observed differences in structural performance. Costs of non-leaded components are currently higher than their leaded counterparts (20% - 30%), but when material costs are evaluated in perspective to the entire cost of the service installation, the overall increase in costs is significantly less (2% - 5%). As more utilities install non-leaded components and the market expands, the material cost differentials should decline.

ALTERNATIVE MATERIALS FOR FITTINGS AND COMPONENTS

Possible approaches to reducing human exposure from lead in brass drinking water fittings and components include: a) use of alternatives to lead to fabricate those components and b) design changes to existing manufacturing procedures (i.e. modifications to current procedures) in order to eliminate lead leaching.

From the standpoint of consumers and regulators, getting the lead out is viewed more favorably, as it would address the lead leaching problem directly and with high certainty. As a general rule with many exceptions, it has been shown that higher lead content in brass results in higher lead leaching to potable water (e.g. Lytle and Schock, 1996). Systematic attempts to identify substitutes for lead resulted in identification of selenium and/or bismuth containing brasses as the most promising solution. Although non-leaded bismuth and selenium-containing brass alloys are the main focus of this study, other materials are discussed as well.

Design changes, including coating and selective leaching of surface lead, might allow use of the existing capital equipment used for manufacturing without expensive modifications or replacements. This alternative is therefore attractive to manufacturers and price sensitive consumers. While this is a relatively painless solution short-term, in the end, the long-term performance of such leaded brass is less certain as the integrity of coatings and surface lead leaching has not been studied in detail. One manufacturer asserted the following about the potential merits of non-leaded cast valves and fittings: "The biggest potential benefit would be a reduction in health hazards to the typical homeowner. While it is difficult to determine a monetary estimate, the regulatory community gives this a very high value. For manufacturers the associated costs may become the cost of staying in business (Showman, 1994)."

Consequently, given the stakes to various stakeholders, the feasibility of alternative non-leaded brass and design changes are described in the sections that follow.

Other Materials for Fabrication (not non-leaded materials)

Faucets made of plastic or faucets with plastic water channels could be considered (Lytle and Schock, 1996). Plastics as an alternative were also discussed in the Organisation for Economic Co-operation and Development (OECD) Workshop on Non-Lead Products and Uses (OECD, 1994). Based on one manufacturer's perspective however, the use of plastic would not provide an economically viable solution for the industry; products would have to be redesigned and the tooling costs for plastic parts could be very high (Anderson, 1994).

Considering options for metallic brasses, besides bismuth and selenium, other elements are known to improve machinability of copper alloys including indium, phosphorous and uranium. In the case of indium, the world supply is rather limited in relation to the new demand that would arise. Phosphorous in high concentrations would react with the water that is present in the green sand molding system. Finally uranium, a radioactive element, can be summarily rejected (Sadayappan et al., 2003).

Phosphorous and indium would probably not offer viable solutions as sole substitutes for lead in brass. However, they could be added to improve mechanical properties of copper alloys in which another substitute for lead (e.g. bismuth) was being used. Laboratory scale studies were conducted at AT&T Bell Laboratories, in which bismuth (and not lead) containing alloys were coupled with phosphorous or indium (among other elements). The machinability of the resultant alloys was equal to that of leaded-alloys (Lytle and Schock, 1996).

Other types of alloys that could be used include silicone cast alloys, lead-free tin bronze, aluminum-bronze and copper-sulfur alloys (OECD, 1994). According to Sadayppan (2005b) copper alloys such as silicon brasses and bronzes offer exceptional casting characteristics in sand and permanent molding operations and they could be considered potential candidates for potable water applications. But the above mentioned are not conventional plumbing alloys, and in order to replace lead-containing materials in drinking water applications they would have to possess acceptable mechanical properties relative to those using lead. Their toxicity would also have to be more thoroughly examined.

In a feasibility study that was performed by the Metals Technology Laboratories of CANMET in 1993, the machinability of tin-bronze, copper-sulfur and silicone brass unleaded alloys was examined. The study concluded that of these alloys copper-sulfur had the best machinability. On the other hand, it was stated that copper-sulfur may not have as much structural strength as the brasses, which could be a drawback in plumbing applications. Silicon brass had moderate machinability and could also have good strength (Sahoo et al 1993).

In addition, for some of these alloys, contamination would have to be rigidly controlled. Silicon contamination of red brasses and tin bronzes can produce excessive leaks in products, whereas in the case of copper-silicon lead is a contaminant of potential concern (Sadayappan, 2005b). Avoiding potential contamination of lead-free alloys during storage and recycling is therefore a concern, and scrap brass might no longer be a suitable source. Scrap brass would have to be used for non drinking water applications (OECD, 1994), unless the removal of lead from scrap was pursued using processes like vacuum distillation (Lytle and Schock, 1996). Lead would also be a contaminant of concern in the case of bismuth copper alloys.

Design Changes

Products could be redesigned in order to reduce internal surface area in contact with drinking water (Lytle and Schock 1996, OECD 1994). Design changes would involve considerable re-tooling costs, depending on the available design options. In the case that new manufacturing methods were to be adopted however, the capital investment could prove to be significant (OECD, 1994).

Chemical treatment has been employed to existing lead components and fittings in order to reduce lead dissolution. The basic approach is to expose the brass surfaces to an aggressive metal extraction solution that removes surface lead from the product (Lytle and Schock, 1996). Research conducted at the University of Florida, where cast products were exposed to a liquid solution of sodium acetate/acetic acid, indicated that such an approach could be feasible (Anderson, 1994). However, based on pilot plant studies that followed, the sodium acetate wash was considered impractical for commercial use (OECD, 1994). In another laboratory study by

Paige and Covino (1992) pretreatment with acetic acid did not significantly reduce the amount of lead leached from commonly used alloys, under the conditions of that specific experiment.

But recent results have been more successful (Ford Meter Box Company, 2005c), and chemical lead removal is currently available in the U.S. and in Europe. There is little doubt that the approach reduces lead leaching over the short term, but the question is whether or not long term leaching is reduced significantly. Research conducted by the Environmental Quality Institute in the 1990's indicated that the various acetate compounds examined were "relatively effective in quickly removing the initial high lead on the inside surface area of a leaded-brass part after sand-casting in a permanent-mold manufacture" (Mass, 2003).

Permanent coatings could also be applied to the internal surface area of a product (OECD 1994, Lytle and Schock 1996, Sadayappan 2005b). The coating could be either organic or inorganic and many of the coating methods have received product approvals from NSF (Sadayappan 2005b) and have been used historically by the water industry for larger diameter valves. In the event that these coatings maintain their integrity long term in smaller diameters applications, and effectively isolate the brass from the water, the lead performance characteristics would be identical to those of unleaded brass.

CHAPTER 3
MAJOR ISSUES AND RESEARCH NEEDS FOR NONLEADED BRASS

INTRODUCTION

Several major research issues and data needs related to non-leaded brass were identified during an Expert Workshop that was convened in December 2005. The group was tasked with identifying and prioritizing the potential issues presented by non-leaded brass, and developing research needs from these issues. Appendix C contains a summary of the Expert Workshop and this chapter summarizes and categorizes the resulting priority data issues and needs.

PRIORITY ISSUES AND RESEARCH NEEDS

A total of 27 priority issues and research needs were identified during the Expert Workshop, and each breakout group prioritized the specific issues they developed. Appendix D presents a listing of these 27 issues/research needs, which break-out group identified the issue, and the priority ranking for each as identified by the breakout group. The breakout groups were formed during the workshop based on each participant's expertise, as shown in Table 3.1.

Table 3.1
Breakout Groups for Expert Workshop

No.	Breakout Group	Participants
1	Utilities	Water utility managers
2	Testing/Professional	Representatives of testing laboratories, manufacturers, and professional organizations
3	Regulatory	U.S. and Canadian federal regulators and a U.S. state regulator
4	Plumbing	Drinking Water Component Manufacturers
5	Manufacturing	Brass alloy manufacturers

While each group identified non-leaded brass research needs specific to their area of expertise, there was also significant overlap in several issues, and in general they could be categorized into the following six major issue areas:

- #1: Leaching, Performance, and Product Acceptance Criteria
- #2: Costs
- #3: Plumbing Code Revisions and Marketplace Enforcement
- #4: Education and Technology Transfer
- #5: Manufacturing Issues
- #6: Regulatory Impact and Implementation

The following sections discuss and summarize the identified issues and research needs for each of these major issue areas.

Major Issue #1 - Leaching, Performance, and Product Acceptance Criteria

A recurring theme during the Expert Workshop was that our understanding of the potential for non-leaded brasses to leach lead, copper and other metals/compounds into potable water is currently limited, as is our understanding of the physical and structural characteristics of non-leaded alloys that impact performance (stress corrosion, cracking, erosion, dezincification, etc.). The importance of understanding both *short-term* and *long-term* field performance of components made using non-leaded brass was identified by the Utility group as their #1 priority data need. Short-term performance was defined as obtaining information on leaching and performance of components utilizing relatively immediate, short-term testing procedures, such as NSF certification testing. Long-term performance was defined as understanding these characteristics (leaching, structural, mechanical, and operability) over the long-term life of the components. This would include how non-leaded components perform in the long-term when compared to lead-containing and "lead-free" components, the expected field performance and corrosion resistance of non-lead brass components, and the service life for various non-lead brass components. Accelerated testing was identified as a possible means towards evaluating long term performance. The ultimate goal identified for both short-term and long-term data gathering was to develop overall performance criteria for non-leaded components based on health and serviceability.

The Regulatory group also identified leaching and physical performance of alloys with low lead-leaching potential (both non-leaded brass and brass subjected to preconditioning, surface treatments, or coatings) as a priority research need. Evaluating various water qualities and treatments, taking into account how component leaching characteristics may change with future changes in water quality characteristics, the nature of the leaching profile over the components' lifetime, and the long-term integrity of surface coatings, plating, and lining processes were specifically mentioned.

The data needs associated with understanding the leaching and performance characteristics of non-leaded alloys and products made from these alloys is directly related to the product approval testing process currently in place (i.e. NSF/ANSI Standard 61, Sections 8 and 9 certification procedures). Identifying key water quality characteristics in drinking waters that impact leaching and performance of non-leaded brass and determining representative drinking water quality characteristics were identified by the testing group as important first steps in possible fine-tuning of the current NSF protocol. Comparing results of current NSF/ANSI Standard 61 Section 8 and Section 9 performance testing of components with results from other studies that have utilized different testing procedures was also discussed.

Major Issue #2 – Costs

Because the installation of non-leaded brass components in the drinking water industry has primarily taken place within the last 5 years, there is a lack of detailed cost information available related to the relative impact of switching to non-leaded in-line components. The potential cost differentials between traditional leaded component and non-leaded component replacement programs may or may not be a significant impact to utilities as they grapple with decisions about materials choices for their distribution system. A closer examination of actual costs incurred by utilities that have installed non-leaded components would be of great value to the industry. In addition, broader costs related to the global impact of raw material supply and recycling will affect both the manufacturing community and the end users of these components and needs further assessment.

Major Issue #3 – Plumbing Code Revisions and Marketplace Enforcement

Updating plumbing codes to incorporate non-leaded brass components was identified as an important issue in the overall standardization of products used in water distribution. Determining what steps would need to be taken to develop the most effective and consistent codes related to non-leaded brass components was identified as an important need that should be implemented quickly. In addition, the need to develop programs to ensure that certified products were being installed and that illegal products were not making their way into the marketplace was also discussed at length. This includes the importance of requiring NSF/ANSI Standard 61 compliance for all products, and validation and third party certification of NSF/ANSI 61 standards. An evaluation of enforcement activities, particularly for illegal suppliers and counterfeit components (including imported products) and conducting point of sale auditing were identified as means of tracking and enforcing the use of required components.

Major Issue #4 – Education and Technology Transfer

The need for development of educational programs and technology transfer opportunities was emphasized by workshop participants. Information sharing between the drinking water community, manufacturers, regulators, and organizations involved in developing and implementing product standards and plumbing codes, quantifying the health benefits of non-leaded components to consumers, and educating stakeholders (utilities, decision makers, regulators, and code enforcers at all levels), were all identified as potential educational programs. The group recommended several possible subject areas they felt should be included as part of this technology transfer, including communicating the costs, benefits and possible constraints of using non-leaded components and understanding the differences in terminology and product definitions ("fixtures" vs. "fittings") currently in use between the different groups. The manufacturing community also stressed the importance of developing and implementing educational opportunities to express the importance of performance criteria to the waterworks industry (e.g. certification based on NSF standards testing).

Major Issue #5 – Manufacturing Issues

Several data needs were identified related to the manufacture of non-leaded alloys and the wide variety of components made from these alloys. The incidental impurities found in recycled materials and the consequent difficulties associated with dissimilar scrap streams from leaded and alternative non-leaded alloys were identified as the highest priority research need by manufacturers. Recycling scrap from manufacturing processes, as well as ensuring that recycled materials are pure, all play into decisions about the costs and applicability of using virgin vs. scrap material. Understanding the global near-term and long-term supply and demand of bismuth, selenium, and other alloying elements used in manufacturing non-leaded brass were also identified as priority issues. Other manufacturing data needs that were identified included the development of alloy machinability standards, assessing the difficulty and ease in processing available alloys, and identifying the manufacturing limitations of alternative materials with respect to both cost and product performance.

Major Issue #6 – Regulatory Impact and Implementation

The primary regulations that impact the use of non-leaded brass components in drinking water systems are the SDWA requirement for allowable lead content for drinking water components, the requirement for faucets and other endpoint components to be independently certified against a voluntary standard, and the federal LCR. The LCR addresses leaching of lead from materials that are used in drinking water distribution and premise piping systems by requiring water systems to test for lead at consumers' taps and mitigate elevated lead levels either through water treatment or by removing and replacing lead source materials from their systems. In addition to the requirement, there are federal requirements for the allowable lead content of solders (<0.2 percent) used to join piping, and for pipe, fittings and components used in drinking water distribution ($\leq$8.0 percent lead content). Priority issues that would need to be addressed in order to incorporate non-leaded brass components into this current regulatory approach were discussed by the regulatory group as well as the entire workshop group.

Since the primary replacements for lead in non-leaded brasses are bismuth and selenium, evaluating the occurrence, chemical characteristics, and toxicology (i.e. potential health effects) of these elements are of concern. Completion of a risk assessment for bismuth (work currently being completed by the USEPA and NSF International) was identified as a vital research need, as was the occurrence and potential health effects of all elements or compounds that may leach from alternatives to leaded brass (P, Si, Bi, Se).

Additional regulatory topics included evaluating regulatory approaches related to non-leaded brass, assisting states in developing and implementing programs related to use of non-leaded brass components, and addressing inconsistencies between federal and state regulations. Also, the workshop participants discussed the need to identify non-regulatory approaches that may influence utility purchasing decisions, with the goal of encouraging the installation of components with very low lead leaching potential.

CHAPTER 4
RESEARCH AGENDA

INTRODUCTION

The primary objective of this project was to present recommended and prioritized research needs related to the use of non-leaded brass components in drinking water, and provide a preliminary structure for the top priority projects to meet those needs. The prioritized research needs, summarized in Chapter 3, were used to develop a research agenda for non-leaded brass that is described in this chapter. This research agenda contains summary descriptions of priority projects categorized into the six major issue areas:

- Leaching, Performance, and Product Acceptance Criteria
- Costs
- Plumbing Code Revisions and Marketplace Enforcement
- Education and Technology Transfer
- Manufacturing Issues
- Regulatory Impact and Implementation.

RESEARCH AGENDA

All of the research needs identified during the Expert Workshop were incorporated into project descriptions, either as a stand-alone project or combined with similar issues into one project. A structure for each of these projects is presented here using a standard format that presents the objective of the research and the approach. These descriptions are not meant to be detailed 'scopes of work' for particular research projects, but rather present a preliminary structure for projects that would meet the identified data needs and can be further developed.

In addition, these projects have been prioritized into "High", "Medium", and "Low" categories. While all of the projects represent identified data needs, there were several issues that were rated highly (i.e. #1 or #2) by more than one break-out group at the Expert Workshop. These issues were important to a broader cross-section of stakeholders represented, and have been given a "High" priority ranking in this research agenda. In addition, projects were listed as "Medium" or "Low" if the issues they represent were identified as lower priority (#3 or lower) and/or only one group identified the issue. Finally, the primary focus of this project is to develop a structure for information gathering that will ultimately be useful to utility providers of drinking water, keeping in mind the interconnection of the regulatory, manufacturing, standards, and utility communities in understanding the issues related to non-leaded brass components. For this reason, those issues of particular interest to utilities were given "High" or "Medium" ratings. A summary of recommended research projects and their priority ranking is displayed in Table 4.1 and described in more detail below.

Table 4.1
Priority Projects for Non-Leaded Brass Drinking Water Components

No.	Major Issue Area	Project	Priority Ranking
1	Leaching, Performance, and Product Acceptance Criteria	1-1. Short Term Testing	High
		1-2. Structural and Mechanical Characteristics	High
		1-3. Long Term Field Testing	High
2	Costs	2-1. Utility Cost Comparisons of Traditional Leaded versus Non-Leaded Component Replacement	High
		2-2. Cost Evaluations of Manufacturing Constraints	High
3	Plumbing Code Revisions and Marketplace Enforcement	3-1. Framework for Plumbing Code Revisions	Medium
		3-2. Mechanisms for Insuring that Certified Products are Installed in the Marketplace	Medium
4	Education and Technology Transfer	4-1. Technology Transfer and Education of Various Stakeholders on the Use of Non-Leaded Brass	High
		4-2. Public Information Programs	Medium
5	Manufacturing Issues	5-1. Effect of Non-leaded Brass Alloys on the Current Recycling Industry	High
		5-2. Global Supply/Demand of Non-leaded Brass Alloy Ingredients	Medium

(continued)

Table 4.1 (Continued)
Priority Projects for Non-Leaded Brass Drinking Water Components

No.	Major Issue Area	Project	Priority Ranking
		5-3. Copper Alloy Machinability Standards	Low
6	Regulatory Impact and Implementation	6-1. Occurrence of Lead Substitute Leachates in Drinking Water	High
		6-2. Identification of Research Needs Related to the Characterization and Toxicology of Lead Substitute Leachates in Brass	High
		6-3. Identification and Evaluation of Various Regulatory Approaches	Low
		6-4. State Regulatory and Program Assistance	Low

MAJOR ISSUE #1: LEACHING, PERFORMANCE AND PRODUCT ACCEPTANCE CRITERIA

Project 1-1: Leaching Characteristics of Low Lead-Leaching Materials and Components

Priority Ranking

High

Objective

The objective of this research is to develop and implement *Short Term Testing Protocols* that will provide information and assurance of the long term leaching characteristics of materials and components used in drinking water distribution that have low lead-leaching potential (i.e. components made using non-leaded brass as well as components made using alternative surface treatment processes such as preconditioning, acid washing and coatings.)

Research Approach

The leaching characteristics of these newer materials that exhibit low lead-leaching potential is currently not well understood. This research would encompass both laboratory and field evaluations of metals leaching characteristics of components with low lead-leaching potential. Existing Standard 61 protocols may be utilized, and/or new testing methods developed such as coupon studies, accelerated testing methods, and/or use of flow through standardized pipe rigs in representative drinking waters around the U.S. Potential partners for this work would include AwwaRF, USEPA, NSF, AWWA, ASDWA, PMI, CDA, other Standard 61 Certification Laboratories, Utilities, and Material and Product Manufacturers.

Specific studies that were identified during the Expert Workshop included:

- Evaluation of Leaching/Metal Release for Low Lead-Leaching Materials and Components Under Real World Conditions
- Impacts of Changes in Raw and Treated Water Quality on Leaching from Materials and Components with Very Low Lead Leaching Potential
- Evaluation of the Impacts of Aggressive Water Qualities on Low Lead-Leaching Materials and Components Using Accelerated Methods.

These are described in more detail below.

Evaluate Leaching/Metal Release for Low Lead-Leaching Materials and Components Under Real World Conditions. The primary objective of this research is to provide a baseline understanding of how materials and components with low lead-leaching potential (non-leaded brass components and components made using acid washed and surface coating procedures) perform with respect to leaching in actual drinking water conditions. This can be divided into three sub-objectives:

1. Identify key water quality characteristics in drinking waters that are aggressive to metal release in leaded, non-leaded, surface coated, and chemically treated brass alloys. This could be accomplished by gathering and evaluating historical information and possibly developing and implementing a field data collection program that would confirm and/or supplement the historical information and provide additional data on parameters that are missing in the historical data,
2. Identify raw and treated water quality characteristics that are representative of drinking water qualities in the U.S. (surface water and ground water sources and both raw and finished (treated) waters) utilizing available data and information collected from state and federal government agencies (e.g. USGS, USEPA, state regulators, others), professional organizations and private research foundations (e.g. AWWA, AwwaRF, ASCE, other), municipal water utilities, and other sources, and
3. Develop and implement protocols to expose components to these identified water quality conditions by obtaining representative low lead-leaching materials and components and testing these products for metal release under both Standard 61 procedures and actual field "simulated" conditions (to be developed), and compare the results.

Impacts of Changes in Raw and Treated Water Quality on Leaching from Materials and Components. The objective of this research would be to:

1. Identify and prioritize the key changes in water that might impact leaching from components made using non-lead alloys are alloys subjected to surface treatments or coatings to reduce lead leaching,
2. Determine the materials in plumbing components that are most likely to be affected, and
3. Determine the contaminants (Se, Bi, Pb, Cu, Sb, organics, and possibly others) likely to be released from the product under a variety of water quality conditions and changes in water quality by developing and implementing short-term, bench scale experiments.

Evaluate the Impacts of Aggressive Water Qualities on Metals Leaching from Low Lead-Leaching Alloys Using Accelerated Methods. Determine long-term leaching characteristics of alloys and components under aggressive water conditions by developing and performing accelerated testing protocols to evaluate metals leaching. Results of accelerated testing must be compared to a subset of actual long-term leaching behavior to establish test method validity.

The product of this research would be technical reports containing results from the specific studies which would be consolidated into an overall report on the leaching characteristics of alloys and components under a variety of water quality conditions. Aggressive water quality characteristics of drinking water, representative characterization of raw and finished drinking water quality in the continental U.S., and descriptions and summaries of the testing protocols implemented and findings should be included. Results could also be used for ongoing re-evaluations of the current NSF protocol.

Project 1-2: Structural and Mechanical Characteristics of Low Lead-Leaching Components

Priority Ranking

High

Objective

The objective of this research is to develop and implement a testing program that will provide information about the structural and mechanical characteristics of materials and components that are used in drinking water distribution and are made using materials with low lead-leaching potential (i.e. either non-leaded brass or alternative processes such as acid washing and surface coatings)

Research Approach

The mechanical and structural characteristics of newer materials with low lead-leaching potential is not well understood. The primary objective of this research is to provide a baseline understanding of both the structural and mechanical properties of new low lead-leaching materials and components, and how those characteristics may change with exposure to actual drinking water conditions. Laboratory protocols should be developed to expose low lead-leaching materials and components to representative drinking water quality and hydraulic conditions followed by identification and/or development of structural and mechanical properties testing procedures to assess basic structural characteristics. Analysis of basic materials properties and how those properties change (such as yield and ultimate stress, strain, hardness, elongation, deformation, ductility, etc.) with exposure to typical drinking water conditions as well evaluations of stress corrosion cracking, erosion corrosion, pitting, dezincification, impacts on the life-cycle of components, and the potential for component failure should be incorporated Testing and evaluation protocols should be used that provide information on new components as well as components that have been exposed to representative drinking water quality conditions for varying lengths of time. Tests should also be designed to evaluate impacts of changes in water quality, and how those changes impact the overall physical performance and operability of the components. Testing protocols could possibly incorporate exposure to representative drinking water qualities using accelerated testing methods followed by evaluation of changes in structural and physical characteristics of low lead-leaching materials and components. Representative water quality conditions, typical changes in water quality that should be incorporated, and accelerated testing protocols can be coordinated with results from Project #1 above.

The product of this research would be technical reports containing results from these studies, which would be consolidated into an overall report on the physical performance of materials and components. Descriptions and summaries of the testing protocols implemented and findings should be included.

Potential partners for this work would include CDA, PMI, Material and Product Manufacturers, NSF, AFS, AwwaRF, USEPA, and materials testing facilities at Universities (i.e. Mechanical, Metallurgical, and/or Materials Science Departments).

Project 1-3: Performance of Installed Non-Leaded Brass Components - Long-term Field Testing

Priority Ranking

High

Objective

Develop a long-term testing and data gathering effort based on a 20- to 50-year horizon to document the performance (leaching, mechanical, operability) of installed components made from non-leaded brass and alternative low lead leaching materials.

Research Approach

The installation of non-leaded brass and alternative low lead-leaching materials is relatively new, and little is known about the long term performance of these materials in the field. This research will identify pertinent information related to leaching and performance of these components, consolidate existing information from utilities that have installed these components, develop and conduct testing at large utilities to characterize leaching and mechanical characteristics and develop a clearinghouse of information contained at a central location (AwwaRF, AWWA) to gather, store, and disseminate the accumulated data at 5-year intervals. The information would be consolidated and summarized in order to determine utility requirements for structural/mechanical/operability characteristics of components. The product would be a 5-year report on performance.

Initial funds would be needed to develop data needs along with on-going financial support of continued data gathering and dissemination of results via a central clearinghouse. Potential partners and/or funding sources for this project would be AwwaRF, AWWA (a possible clearinghouse), CDA, and utilities.

MAJOR ISSUE #2– COSTS

Project 2-1: Utility Cost Comparisons of Traditional Leaded versus Non-Leaded Component Replacement

Priority Ranking

High

Objective

Develop utility cost information for installation of traditional leaded component versus non-leaded components in utility replacement programs.

Research Approach

Utilities have a need to understand the cost impacts of installing non-leaded components in their system. The research approach will involve developing cost information needs related to distribution system component replacement programs (i.e. unit costs for all in-line devices, both leaded and non-leaded components; combined costs for all labor and materials used; life cycle of components, labor cost differentials, etc.). Utilities will be identified and surveyed to gather the data which will be analyzed and summarized. Information to be evaluated should include utility costs associated with switching from typical leaded components to non-leaded components. This should include per unit costs for various non-leaded in-line components and overall program costs such as costs for additional materials that must also be replaced (fittings, etc.), requirements for additional or different tools or methods for installing non-leaded brass, and labor costs. Comparisons between traditional leaded component replacement versus non-leaded component replacement should be included and an assessment of the relative impact of using non-leaded materials to the overall program costs for replacement of in-line components should be calculated (i.e. component costs as a percent of total costs). This assessment should also take into account the potential changes in the differential costs between traditional leaded versus non-leaded components as the market for non-leaded components becomes larger. Utilities should be able to use this information for planning the costs associated with their conversion to non-leaded components. The product would be a brief summary report. Potential partners and/or funding sources would likely include AwwaRF, utilities, ASDWA, and AWWA.

Project 2-2: Cost Evaluations of Manufacturing Constraints

Priority Ranking

High

Objective

Evaluate the broad cost impacts of raw material supply, recycling, and other constraints on the manufacturing of non-leaded alloys and components.

Research Approach

Manufacturers need to have a more precise understanding of the cost implications of manufacturing non-leaded brass alloys and components. This research will identify the major cost impacts related to manufacture of non-leaded brass materials and materials with low lead-leaching potential. Cost impacts that should be considered include the impacts of changes in the supply of non-leaded brass alloy ingredients, recycling of non-leaded brass components, costs associated with foundry scale up requirements and design and manufacturing process modifications, and costs related to manufacturing implications of using other alternative non-leaded materials and or processes. The research will essentially be a global impact analysis/market study of the cost impacts of these factors. The product of this work would include a technical report summarizing these cost factors. Potential funding sources for these projects would include the CDA, the American Foundry Association, the Plumbing Manufacturer's Institute, and alloy and component manufacturers.

MAJOR ISSUE AREA # 3 – PLUMBING CODE REVISIONS AND MARKETPLACE ENFORCEMENT

Project 3-1: Framework for Plumbing Code Revisions

Priority Ranking

Medium

Objective

Develop a framework for implementing revisions to plumbing codes related to the use of drinking water components with low lead-leaching potential.

Research Approach

The issue is that there is not a clear understanding of how current plumbing codes address drinking water components with low lead-leaching potential and how codes may need to be revised to incorporate these materials. The research approach should contain two major parts:

1. Gathering, evaluating, and summarizing basic background information about current international, national, and state plumbing codes that apply to the use of drinking water components with low lead-leaching potential. This could include an assessment of which NSF/ANSI Standard 61 sections or other applicable standard (i.e. ASME) are referenced in various codes and identifying time-frames and schedules for implementing changes or revisions to plumbing codes.
2. Recommending revisions for future versions of plumbing codes as they relate to products with low lead-leaching potential, and developing a framework for implementing those revisions. The framework should outline the necessary steps that need to be taken to implement those revisions, the entities that will need to be involved, and timeframes for implementation.

This approach could be implemented by convening a working Task Group of plumbing code developers, regulators, manufacturers, and NSF representatives. The product of this research would be a Task Group report summarizing the current status of national plumbing codes (with respect to non-leaded brass); the relationship between national and local code development and utility specification procedures; and any other historical and current information gathered that would support or refute possible revisions to the codes. Potential partners and/or funding sources for this work would be NSF, the PMI, and the International Association of Plumbing and Mechanical Officials (IAPMO).

Project 3-2: Mechanisms for Insuring that Certified Products are Installed in the Marketplace

Priority Ranking

Medium

Objective

Identify mechanisms for identifying, controlling, and enforcing the use of certified drinking water components.

Research Approach

The issue is that there are many components being sold and installed that do not meet standards or certification requirements. This research will identify mechanisms that could effectively assist in controlling and enforcing the use of certified drinking water components in the marketplace. This could include:

- An evaluation of the status of validation and third party certification of NSF/ANSI 61 standards for products used in water distribution,
- Legal mechanisms for controlling drinking water components and enforcing illegal installation activities,
- Identifying and developing programs designed to trace counterfeit products (forgeries of original products introduced into the marketplace that may contain different metals contents than the original),
- Enforcing non-permitted installation,
- Monitoring purchase/transactions at the consumer level, and
- Tracing plumber (customer) vs. utility installed components.

Specific program elements may include the need for inspectors, completing accounting reviews to determine if specified items were actually purchased, and/or tracing point-of-sale paper-work followed by testing the actual faucets that were installed. The product of this research would be a technical report containing recommended legal and enforcement mechanisms that can be implemented to ensure that certified drinking water components are installed in the U.S.

Potential partners/funding sources for this research would include the International Association of Building and Mechanical Officials, the Plumbing Manufacturer's Institute, the USEPA, and State and Local plumbing code agencies.

MAJOR ISSUE AREA # 4 – EDUCATION AND TECHNOLOGY TRANSFER

Project 4-1: Technology Transfer Programs and Education of Various Stakeholders on the Use of Non-Leaded Brass

Priority Ranking

High

Objective

Gather information and develop programs to educate various stakeholders (utility, decision makers, regulators, standards developers, code enforcers) on issues related to the use of non-leaded brass.

Research Approach

There is currently a lack of consistent understanding amongst various stakeholders of the issues related to non-leaded brass. This research will develop educational tools, such as technology transfers and workshops to encourage education and cross-pollination of information with the public, i.e. decision makers and various stakeholders (utility, decision makers, regulators, standards developers, code enforcers) at all levels. Collect and summarize supporting data to be presented using historical and current information and organize opportunities for data sharing between utilities, industry, standards developers, and regulators. Determine the most effective combination of technology transfer options and tools to use, which could include resources provided by AWWA, AwwaRF, USEPA, the Centers for Disease Control (USCDC), the National Institutes of Health (NIH), the Agency for Toxic Substances and Disease Registry (ASTDR), and the Association of State Drinking Water Administrators (ASDWA). Content could include:

- Information on health impacts of use of various materials by USCDC, such as correlation of lead in water to blood lead levels,
- Utility data sharing of experiences related to non-leaded brass use,
- Links between the AWWA standards and NSF/ANSI Standard 61,
- Regulatory and certification approaches for non-leaded brass components, and
- Incorporation of non-leaded materials into Green Design Standards.

The product of this work would be technology transfer seminars/workshop materials including presentation papers and additional information. The potential partners and/or funding sources for the project include AWWA, AwwaRF, CDA, PMI, IAPMO, and utilities.

Project 4-2: Public Information Programs

Priority Ranking

High

Objective

Gather information and develop programs that utilities can use to educate their customers on the use of drinking water components with low lead-leaching potential.

Research Approach

The issue is that the general public is not informed about the advantages and availability of components with low lead-leaching potential. This research project will develop educational tools and public information programs that can be used to educate the public on the use of drinking water components with low lead-leaching potential. Information should be collected and summarized on:

- Health impacts of using components with low lead-leaching potential in order to inform consumers of the health benefit of low lead-leaching components.
- Cost impacts of replacing current leaded components with low lead-leaching components
- Timeframes for replacement of utility components

These materials should contain general information that can be tailored to each utilities specific plans for potential replacement of leaded components with components with low lead-leaching potential. Utilities should be able to fine-tune the materials to develop individual public information programs that apply to their specific operations and maintenance plans for incorporating components with low lead-leaching potential into their systems. The focus of these public information materials will be to inform customers of utility programs related to low lead-leaching components, but could also contain information on the availability of premise piping components with low lead-leaching potential and the importance of customer purchasing of certified premise piping products.

The product of this work would be public information materials including presentations, pamphlets, and brochures. The potential partners and/or funding sources for the project include AWWA, AwwaRF, utilities, CDA, Plumbing Manufacturer's Institutes, and IAPMO.

MAJOR ISSUE AREA # 5 – MANUFACTURING ISSUES

Project 5-1: Effect of Non-leaded Brass Alloys on the Current Recycling Industry

Priority Ranking

High

Objective

Evaluate the current and projected future effect that manufacture of non-leaded brass alloys may have on the recycling industry with respect to copper alloy ingredients (including copper, bismuth, selenium, etc). Assess the impacts of these recycling effects on the manufacture of non-leaded alloys and components, particularly the effect that bismuth may have on recycling and the ultimate costs associated with manufacturing of non-leaded alloys and components.

Research Approach

The manufacturing community does not have an adequate understanding of how the manufacture of non-leaded brass alloys and components will affect (and be affected by) the recycling market. This research will conduct a market study to identify and quantify major issues related to recycling of copper alloys containing bismuth and selenium. The study should be developed so that key information is gathered, analyzed, and presented. This could include conducting a market research survey that quantifies future recycling trends, costs associated with recycling, availability of recyclers who will accept bismuth containing materials, and how other industries may handle dissimilar scrap streams. The analyses should identify what factors related to non-leaded alloys would significantly hamper the future of recycling, the effect that bismuth may have on recycling of copper alloys, and the potential that products may be sent to a landfill instead of recycled. The study should include estimates of the size of the non-leaded cast alloy market and costs associated with sending scrap back to refiners. Possible sources of information include Institute of Scrap Recycling Industries (ISRI) and bismuth refiners. The product would be a Market Study completed using the combined resources of CDA, PMI, AFS, and manufacturing companies.

Project 5-2: Global Supply/Demand of Non-leaded Brass Alloy Ingredients

Priority Ranking

Medium

Objective

Define trends, issues, and associated costs related to the supply of non-leaded brass alloy ingredients.

Research Approach

The manufacturing community needs a clearer understanding of the global supply and demand of ingredients required to manufacture non-leaded brass alloys. This research will develop and conduct a study that quantifies the current usage and forecasted growth of cast and wrought leaded fittings in North America. This would include a determination of the amount of raw material needed to replace the lead (Se, Si, Bi, Cu) in alloys, projections of the worldwide supply of raw materials for all uses, and determination of price points or other triggers that would impact the amount of material available. A timeline of material availability that includes the impacts of recycling streams should be projected and near and long-term availability and pricing of bismuth, selenium, silicon, antimony, and copper should be included. The product would be a Market Study that includes a Supply/Risk/Economic Model. Potential partners and/or funding sources for the project include CDA, the Copper and Brass Fabricator's Council, the Non-ferrous Founder's Society (NFFS), AFS, and USGS.

Project 5-3: Copper Alloy Machinability Standards

Priority Ranking

Low

Objective

Develop a useful and standard machinability index for wrought and cast copper alloys.

Research Approach

There is currently no standard machinability index for non-leaded brass alloys. Having a standard index would provide for consistency in the manufacture and tooling of non-leaded brass components. This research will develop and implement an applied research plan aimed at identifying a standard machinability index that can be used by manufacturers of non-leaded brass components. The responsibility for this undertaking should reside within the copper industry and be coordinated with material users that conduct machining and other industry groups (e.g. PMI). The research should identify and evaluate machine tool characteristics (geometry, tool wear, tool life, cutting force, power requirements, cutting temperatures, cutting speed and force, feed rate, etc) and material composition and could incorporate pilot and/or field studies. Results should include a definition of machinability, descriptions of the methods of testing for machinability, and an independent third party verification to compare alloys head-to-head. The product would be a Machinability Standard/Index that would be developed and funded through the joint efforts of CDA, PMI, and manufacturers.

MAJOR ISSUE AREA #6 – REGULATORY IMPACT AND IMPLEMENTATION

Regulatory review and analysis of potential drinking water regulations generally involves evaluation of five elements: health effects, analytical and treatment feasibility, implementation-related issues, occurrence and exposure, and economic impacts. During the workshop, attendees identified several regulatory research needs related to lead substitute leachates, including occurrence, health effects, and identification and evaluation of regulatory strategies. Taken together, these projects constitute the beginnings of a Regulatory Impact Analysis for non-leaded brass leachates. These projects are summarized below.

Project 6-1. Occurrence of Lead Substitute Leachates (Se, Bi, P, Si, etc..) in Drinking Water

Priority Ranking

High

Objective

The objective of this research is to estimate the occurrence of lead substitute leachates in drinking water in the U.S.

Research Approach

Potential future regulation of lead substitute leachates should be based on a thorough understanding of their occurrence in the environment. For this project, a research approach should be developed that will provide estimates of occurrence of lead substitute leachates in drinking water. This approach should include:

1. Identification of possible lead substitute leachates that would occur in distributed drinking water due to the use of components with low lead-leaching potential. Likely parameters would include Se, Bi, P, and Si, however, an evaluation and justification for the inclusion of additional parameters should be completed.
2. Identification and evaluation of historical data on the occurrence of these identified parameters in drinking water. It is likely that the historical occurrence data for these parameters will be sparse; however, data could be gathered by completing a literature review and/or a survey of drinking water utilities. Other potential sources of historical data may include the Safe Drinking Water Information System (SDWIS), the USGS's National Water Information System (NWIS), USEPA databases such as the Unregulated Contaminant Information System (URCIS), and other privately- and publicly-available data sources and the National Drinking Water Contaminant Occurrence Database (NCOD).
3. Development and implementation of a program to collect new data on the presence of the identified lead substitute leachate parameters. The data collection plan should be geographically and spatially diverse and identify a representative cross section of drinking water sources, treatment, and low lead-leaching drinking water components. Data should be collected from the source and treated water to provide baseline levels of these parameters in drinking water, and from the distribution system and taps to identify their occurrence due to leaching from components with low lead-leaching potential.

It may be impossible to collect the amount of data necessary to accurately assess occurrence, however the data collection plan should attempt to collect the most complete information possible with the funds available. The data should be statistically evaluated to estimate occurrence and if possible, any trends in the occurrence of these parameters. These results should be compared with existing data and knowledge of lead and used to evaluate existing and possible future regulatory approaches. The product of this research would be an Occurrence Document that can be used as part of a larger regulatory impact anaylsis. This research would be funded and managed by the USEPA, with additional partners such as AwwaRF, AWWA, CDA, and NSF International.

Project 6-2: Identification of Research Needs Related to the Characterization and Toxicology of Lead Substitutes in Brass (Si, Bi, P, Se, etc).

Priority Ranking

High

Objective

Identify research needs for properly characterizing the chemistry, speciation, and toxicology of lead substitute leachates (Se, Bi, P, Si, etc.).

Research Approach

The USEPA needs to understand the toxicology and the chemical characteristics of lead substitutes in brass in order to evaluate regulatory approaches. This project will gather and evaluate current and historical data on the chemical and toxicological characteristics of lead substitute leachates such as selenium bismuth, phosphorus, and silica, and develop a research plan that will further characterize the chemical and toxicological attributes of these parameters. The plan should utilize and summarize current toxicological evaluations (such as ongoing health effects studies by the USEPA and NSF International), identify and gather any additional historical information needed to supplement these studies, identify where data gaps may exist with respect to the characterization of these lead substitute leachates (including their radioactivity, oxidation states, and chemistry/speciation data gaps), determine if those gaps can be resolved through experimentation, and identify what experiments would be most appropriate.

The product of this research would be a technical report summarizing the current state of knowledge with respect to the chemical and toxicological characteristics of lead substitute leachates, identifying research gaps, and recommending a research plan to fill those gaps. Specific projects, costs, and timelines should be identified in the research plan.

Project 6-3: Identification and Evaluation of Various Regulatory Approaches

Priority Ranking

Medium

Objective

Develop and evaluate regulatory strategies to reduce lead leaching potential from plumbing devices.

Research Approach

Develop regulatory strategies and conduct a systematic evaluation of the practical benefits and limitations of each strategy. Factors to consider include:

1. cost (manufacturers and increased product price)
2. time to achieve regulatory change
3. time to achieve beneficial outcome of regulatory change
4. scope of products covered
5. ancillary benefits (e.g., regulating at point of sale might reduce exposure in private well systems, reduction of lead in recycle streams)
6. regulatory staff required
7. required structure and resources to enforce
8. can it be enforced
   a. resources required
   b. legal basis

For each of the above, qualitatively and quantitatively try to determine what has been done previously and its effectiveness, and consider new approaches that might complement existing approaches. Develop a clear set of actions, the timing and combination of which, will best achieve lower lead leaching potential from plumbing devices. This could include the need to re-write regulations and the impact of other regulations on non-leaded brass (simultaneous compliance issues). Also, the impact that replacement of kitchen faucets with very low lead leaching components may have on current regulatory compliance (i.e. 90 percent lead levels) and overall public health implications if other components (bathroom faucets, fountains) are not replaced. Finally, evaluating the pros and cons of performance based standards (i.e. NSF/ANSI Standard 61) vs. a regulatory approach that prescribes allowable lead content of materials should also be included in the analyses.

The product would be a preliminary Regulatory Impact Analysis Report completed by the USEPA and State Drinking Water Regulatory Agencies

Project 6-4: State Regulatory and Program Assistance

Priority Ranking

Low

Objective

Identify steps that the federal government can take to assist states in developing and implementing programs related to the use of non-leaded brass components in drinking water systems.

Research Approach

The issue is that there are currently many different state approaches for encouraging the use of non-leaded brass components and regulating their use. This project will convene a working Task Group of federal and state regulators and other stakeholders (code enforcers, manufacturers, industry groups) to identify inconsistencies between various state regulations, to identify potential programs that can be implemented at the state level to encourage the use of low lead-leaching components, and to develop a framework for implementing these programs. This could include identifying non-regulatory approaches, i.e. incentives that can be offered to the drinking water industry to encourage the utilization of non-leaded components by both consumers and utilities and mechanisms for enforcing illegal leaded brass component installation. The product of this research would be a Task Group Report that:

1. Identifies current inconsistencies in state regulations and codes related to components with low lead-leaching potential,
2. Recommends steps that can be taken to correct those inconsistencies in state regulations and codes, and
3. Identifies program elements and timeframes for several non-regulatory approaches that could be implemented at the state level to encourage the use of low lead-leaching components by the waterworks industry and consumers.

Potential partners for this work would be the USEPA, State Regulatory Agencies, and the IAPMO.

CONCLUSIONS

A Research Agenda for non-leaded brass was developed that includes sixteen priority projects. A preliminary structure for each project was identified that can be further developed into 'scopes of work' for specific research projects. Those projects identified as 'High" priority in the research agenda included:

Major Issue Area	Project
Leaching, Performance and Product Acceptance Criteria	• Short-Term Testing. • Structural and Mechanical Characteristics. • Long-Term Field Testing.
Costs	• Utility Cost Comparisons of Traditional Leaded versus Non - Leaded Component Replacement. • Cost Evaluations of Manufacturing Constraints.
Education and Technology Transfer	• Technology Transfer and Education of Various Stakeholders on the Use of Non-Leaded Brass.
Manufacturing Issues	• Effect of Non-leaded Brass Alloys on the Current Recycling Industry.
Regulatory Impact and Implementation	• Occurrence of Lead Substitute Leachates in Drinking Water. • Identification of Research Needs Related to the Characterization and Toxicology of Lead

Major Issue Area	Project
	Substitute Leachates in Brass.

Those projects identified as 'Medium' priority in the research agenda included:

Major Issue Area	Project
Plumbing Code Revisions and Market place Enforcement	• Framework for Plumbing Code Revisions. • Mechanisms for Insuring that Certified Products are Installed in the Marketplace.
Education and Technology Transfer	• Public Information Programs.
Manufacturing Issues	• Global Supply/Demand of Non-leaded Brass Alloy Ingredients.

Projects identified as 'Low' priority included:

Major Issue Area	Project
Manufacturing Issues	• Copper Alloy Machinability Standards.
Regulatory Impact and Implementation	• Identification and Evaluation of Various Regulatory Approaches. • State Regulatory and Program Assistance.

This research agenda was developed to fill in knowledge gaps related to issues surrounding non-leaded brass materials used in potable water systems, and can be used as a framework for organizing and prioritizing research funding decisions.

APPENDIX A
NSF/ANSI STANDARD 61 SECTIONS 4, 8, AND 9 PRODUCT EVALUATION OVERVIEW

NSF/ANSI STANDARD 61: DRINKING WATER SYSTEM COMPONENTS – HEALTH EFFECTS

Section 4, 8 and 9 Product Evaluation Overview (Greiner 2006)

INTRODUCTION

In response to a competitive request for proposals from the U.S. Environmental Protection Agency (USEPA), a Consortium led by NSF International (NSF) agreed to develop voluntary third-party consensus standards and a certification program for all direct and indirect drinking water additives. Other members of the Consortium included the American Water Works Association Research Foundation, the American Water Works Association, the Association of State Drinking Water Administrators, and the Conference of State Health and Environmental Managers. (COSHEM has since become inactive as an organization.) Each organization was represented on a committee along with balanced representation from the regulatory community, manufacturing industry, and water utilities. This Joint Committee on Drinking Water Additives continues to maintain the standards today,

Two standards for additives products were developed. NSF/ANSI Standard 60 – *Drinking water treatment chemicals — Health effects* covers many of the water treatment chemicals, also known as direct additives. NSF/ANSI Standard 61 – *Drinking water system components — Health effects* covers all indirect additives products and materials. Testing to determine the potential of a product to impart taste and/or odor to drinking water is not included in these Standards.

NSF/ANSI 61 was developed to establish minimum requirements for the control of potential adverse human health effects from products that contact drinking water. It does not attempt to include product performance requirements that are currently addressed in the USEPA's ETV program, other NSF standards, or other voluntary consensus standards established by such organizations as the American Water Works Association or the American Society for Testing and Materials.

NSF/ANSI 61, and subsequent product certification against it, has replaced the USEPA Additives Advisory Program for drinking water system components. USEPA terminated its advisory role in April 1990. For more information with regard to USEPA's actions, refer to the July 7, 1988 *Federal Register* (53FR25586).

These Standards are intended for use by certifying organizations, utilities, regulatory agencies, and/or manufacturers as a basis of providing assurances that adequate health protection exists for covered products. Product certification issues, including frequency of testing and requirements for follow-up testing, evaluation, enforcement, and other policy issues, are not currently addressed by the Standard.

LIMITATIONS OF OVERVIEW

This document has been written to provide an overview on the evaluation of products to ANSI/NSF Standard 61. The standard in its entirety should be used when evaluating products for compliance. This document has been limited to cold-water evaluations on products covered in Section 4, Section 8, and Section 9 of the Standard. (Products typically used in and around homes, such as in-line and end-point devices less the 4" nominal ID and connective pipe and fittings.) Other options included in the standard but not presented in this overview include the following.

- Evaluation of products to hot water conditions.
- Options for performing 90-Day "overtime" analyses that establish extraction decay rates of contaminants. (Establishment of extraction decay rates enables separate health effect assessments of the concerns of acute short-term exposures and chronic long-term exposures. This option is not viable for many contaminants including lead.)

SCOPE OF STANDARD

Standard 61 covers products and materials that come into contact with: drinking water, drinking water treatment chemicals, or both. The focus of the Standard is evaluation of contaminants or impurities imparted indirectly to drinking water. The products and materials covered include, but are not limited to, process media (carbon, sand, etc.), protective materials (coatings, linings, liners, etc.), joining and sealing materials (solvent cements, welding materials, gaskets, etc.), pipes and related products (pipes, tanks, fittings, etc.), mechanical devices used in treatment/transmission/distribution systems (valves, chlorinators, separation membranes, etc.), and mechanical plumbing devices (faucets, endpoint control valves, etc.). Covered products are grouped into sections:

SECTION [1]	Example Products
4	Pipe, fittings, well casing, riser tube
5	Coatings, liners, cements
6	Gaskets, adhesives, lubricants
7	Sand, gravel, ion exchange resins
8	Valves, meters, process equipment
9	Faucets, flex plumbing connectors, stop valves, fountains

[1] Sections 1, 2 and 3 contain requirements common to all the evaluation of all products.

Point-of-use and point-of-entry drinking water treatment devices are not covered by the scope of Standard 61 as they are covered by other NSF standards. Fire hydrants have also been specifically excluded from the scope of Standard 61.

STANDARD EVALUATION PROCESS

NSF/ANSI Standard 61 is concerned about the potential extraction of any chemical, not just lead. The standard follows a risk-assessment style approach that enables:

- Identification of substances contributed to drinking water by products or materials.
- Estimation of human exposure to the substance.
- Evaluation of the potential health safety concerns presented by the exposure.
- Estimation of the acceptable risk associated with the use of the product in drinking water.

A common evaluation process applies to all products. The process steps are:

Information collection: As Standard 61 is concerned about any potential contaminant that may leach from products the standard requires that formulation information on all water contact material be provided in order to identify potential extractants. This includes detailed information from manufacturers and their suppliers on composition, known or suspected impurities, and manufacturing processes for all wetted components in products submitted for evaluation.

Identification of potential extractants: Two items establish test requirements. First, Table 3.1 of the standard identifies the minimum battery of testing required and is driven by the generic type of materials used in the water contact parts. Secondly, any manufacturer specific or formulation specific analytes of concern determined during the review of the information collected are added to the test battery.

Laboratory testing: **Laboratory testing provides a mechanism to establish if any of the potential extractants of concern extract under standardized testing conditions. The testing protocols are all performed as "in-vessel" or "in-product" exposures through a simple "dump and fill" exposure style. They are not routinely performed plumbed-in or under pressure. The laboratory testing has three components.**

- Sample preparation
- Conditioning (water not analyzed)
- Exposure (water analyzed)

Controls are used throughout the evaluation process to assure that contaminants can be confirmed as coming from the product rather than introduced as an artifact of the evaluation.

Normalization of results to potential at-the-tap values: **Normalization of the laboratory findings is performed to translate laboratory extraction testing to potential at-the-tap concentrations based on field use assumptions of the product type.**

Evaluation of normalized extractant values to drinking water acceptance criteria. The drinking water acceptance criteria are established through Annex A of the standard. The criteria for over 600 chemicals have been established and are listed in Annexes D and E. The requirements are based on regulated levels and health advisories where they exist and based on available toxicology data where they don't. USEPA and Health Canada representatives have established harmonized acceptance criterion for the standard where were differences exist between a USEPA MCL (maximum contaminant level) and a Health Canada MAC (maximum allowable concentration).

Standard 61 contains four types of acceptance criteria

Criteria	Description	Comments
STEL	• Short term exposure level	These are only used to assess the potential health effects from the short-term exposure portion of 90-day evaluations. This is most commonly used for volatile and semi-volatile compounds when their extractions are known or suspected to rapidly decay.
TAC	• Total allowable concentration • Same as regulated levels when exist	These are only used when evaluating the results of testing that have been normalized to fields use assumptions under potential static dwell conditions. This is the criteria applied to the results from in-line device exposures and connective pipe and fittings under 4" nominal ID
SPAC	• Single product allowable conc. • Typically 1/10th of TAC	These are used for Section 9 products for all extractants other than lead. They are also used for larger Section 4 and Section 8 products ($\geq$ 4" nominal ID) when the results of testing have been normalized to "flowing" field use assumptions.
Q	• Lead test statistic ("Q factor") • 11 µg	This is the maximum allowable lead extraction from Section 9 devices.

The acceptance criteria for lead under the standard.

Criteria	Acceptance Level	When used
STEL		Not an option for lead evaluations
TAC	15 µg/L	Used on the Sec 4 & 8 products < 4" nominal ID
SPAC	0.5 µg/L	Used on the Sec 4 & 8 products $\geq$ 4" nominal ID
Q	11 µg	Used on all Sec 9 devices

The process steps for information collection and identification of potential extractants are the same for all product types. No additional details on these have been included in this overview. Steps for laboratory testing, normalization, and evaluation of test results are product specific. Additional details on these steps are in the sections that follow.

PRODUCT TESTING, NORMALIZATION, AND RESULT EVALUATION

Section 4 and Section 8 products

Laboratory testing

The product testing (preparation, conditioning, and exposure) protocols for Section 4 and Section 8 products are nearly identical and are noted in the table below as are the acceptance criteria. When testing for metals extractants, separate products are required to be exposed to both pH 5 and 10 exposure waters. When non-metal analysis is required, a pH 8 water is used. All exposure waters are required to be made fresh daily.

Step	Section 4 Products (< 4" Nominal ID)	Section 8 Products (< 4" Nominal ID)
Sample Preparation	Rinse product with tap water to remove any debris or contamination that occurred during shipping and handling of the product followed rinse with reagent water [1]	Same
Conditioning	Expose for 16 days with 12 water changes @ 23° C (All waters are discarded)	Same
Exposure	Expose for 16 hours @ 23° C (Water analyzed)	Expose for 12 to 16 hours @ 23° C (Water analyzed)
Normalization	See normalization section that follows.	See normalization section that follows.
Acceptance Criteria	TAC	Same

[1] The term "reagent water is used in the Standard to refer to the laboratory's high purity water used as the base that all test solutions are made from. It can be produced through distillation, reverse osmosis, ion exchange, or equivalent processes and must have the following properties:

- Electrical resistivity, minimum 18 MΩ-cm at 25° C
- Total organic carbon, maximum 0.1 mg/L

Normalization

Normalization provides a mechanism to estimate potential "at-the-tap" concentrations based on the level of contaminants identified during laboratory testing. The following equation shows the variables used during normalization of Section 4 and Section 8 products with a nominal ID < 4".

$$NF = N1 \times N3$$

$$N1 = \frac{SA_F}{SA_L} \times \frac{V_L}{V_{F(static)}}$$

where:

SA_F = surface area exposed in the field;
SA_L = surface area exposed in the laboratory;
V_L = volume of extraction water used in the laboratory;
$V_{F(static)}$ = volume of water the product is exposed to in the field for the static condition (for devices that hold less than 1-L, the $V_{F(static)}$ value used is 1 L)
N3 = 0.33 (N3 is only used during the normalization of test results from in-line devices covered in Section 8)

EVALUATION OF NORMALIZED TEST RESULTS

The normalized concentration must not exceed the "Total allowable concentration" (TAC) drinking water acceptance criteria where listed in Annex D or E of the Standard. If acceptance criteria is not listed in either annex D or E then a TAC criterion is required to be established under Annex A of the Standard before the evaluation of the acceptability of the results can be completed.

Product testing, normalization, and result evaluation - Section 9 products

Laboratory testing

The evaluation of products under Section 9 requires use of only one exposure water, whether looking for metals, non-metals or both. The exposure water used is pH 8 solution with an alkalinity of 500 mg/L, dissolved inorganic carbon of 120 mg/L, and 2 ppm of free chlorine. As with all exposure waters used under the standard, the extraction water is required to be made fresh daily.

a) Step	Section 9 Products
Sample Preparation	Flush for 15 minutes with tap water followed by a rinse with 3 unit volumes of reagent water [1]
Conditioning	Rinse with 3 unit volumes of reagent water. The units are then filled with test water and held until the beginning of the exposure series (All waters are discarded)
Exposure	A 19-day exposure protocol is performed according to Figure B1 (attached). For lead analysis testing is required to be performed in triplicate as a minimum and the waters from the 16-hour exposures from 9 of the exposures (Days 3, 4, 5, 10, 11, 12, 17, 18, and 19) are collected for analysis. For non-lead analysis, the testing of only one unit is required and only the 16-hour sample from Day 19 is analyzed.
Normalization	See normalization equation that follows.
Acceptance Criteria	SPAC for non-lead contaminants. Q of 11 µg for lead (calculation below)

[1] The term "reagent water is used in the Standard to refer to the laboratory's high purity water used as the base that all test solutions are made from. It can be produced through distillation, reverse osmosis, ion exchange, or equivalent processes and must have the following properties:

- Electrical resistivity, minimum 18 MΩ-cm at 25° C
- Total organic carbon, maximum 0.1 mg/L

Normalization

The equation used to normalize Section 9 device test results is similar to that used with the results of testing on Section 4 and Section 8 products as described above.

$$NF = N1 * CMV$$

The variables in the N1 factor are the same. The N3 factor is not used. The $V_{F(static)}$ value for most products is 1 L. For commercial kitchen devices it is 18.9L. A "cold mix volume" (CMV) factor is also used which is calculated as the devices ("cold water volume") / ("total device volume").

Evaluation of non-lead results

The normalized concentration must not exceed the "single product allowable concentration" (SPAC) drinking water acceptance criteria where listed in Annex D or E of the Standard. If acceptance criteria is not listed in either annex D or E then a SPAC criterion is required to be established under Annex A of the Standard before the evaluation of the acceptability of the results can be completed.

Evaluation of lead results

The analytical protocol under Section 9 generates nine measured lead dosages (on Days 3, 4, 5, 10, 11, 12, 17, 18, and 19) leached from each of the products sampled from a particular product line. The number of products tested is defined as "n." The test data can be described as values of D_{ij} (ith product measured on the jth day).

$$Y_{ij} = \ln(D_{ij})$$

$$Y_i = \frac{(Y_{i3} + Y_{i4} + Y_{i5} + Y_{i10} + Y_{i11} + Y_{i12} + Y_{i17} + Y_{i18} + Y_{i19})}{9}$$

$$D_i = e^{Yi}$$

$$\text{Log - dosage mean, } \overline{Y} = \frac{\sum_{i=1}^{n} Y_i}{n}$$

$$\text{Log-dosage standard deviation, } S = \sqrt{\frac{\sum_{i=1}^{n} (Y_i - \overline{Y})^2}{(n-1)}}$$

The test statistic, Q shall be determined as:

$$Q = e^{\overline{Y}} \cdot e^{k_1 \cdot S}$$

The value of k_1 depends upon the sample size. When the minimum sample size of three is used, $k_1 = 2.60281$. As the number of samples increases, the k_1 value decreases. Table B11 of the Standard contains the value of k_1 for sample sizes up to 50. When the maximum sample size of 50 is used, $k_1 = 0.89361$. The acceptability of the product line depends upon the value of the test statistic, where:

- Case I: If $Q \leq 11$ µg, the product line has tested as acceptable; or
- Case II: If $Q > 11$µg, the product line has tested as unacceptable.

PRODUCT CERTIFICATIONS, LISTINGS, AND FOLLOW-UP

Product certification provides independent verification that products fully comply with the requirements of the standard. At NSF, after the product is listed each certified manufacturing site receives unannounced audit(s) by our field staff. During this visit, the NSF representative again verifies that the manufacturing process of the certified product does not depart from the original formulation or design. NSF reviews the manufacturing process and quality assurance records and collects a chain-of- custody monitoring sample of the certified product for follow-up testing in our laboratories to assure that there have been no changes to the product that could create adverse health effects to consumers.

Figure B1 – Exposure sequence for mechanical plumbing device (NSF/ANSI 61 – 2005)

	Fri	Sat	Sun	Mon	Tue	Wed	Thu	Fri	Sat	Sun	Mon	Tue	Wed	Thu	Fri	Sat	Sun	Mon	Tue	Wed	Thu	Fri
Test Day				**1**	**2**	**3**	**4**	**5**	**6**	**7**	**8**	**9**	**10**	**11**	**12**	**13**	**14**	**15**	**16**	**17**	**18**	**19**
						C	C	C					C	C	C					C	C	C
	W/C			2	2	2	2	2			2	2	2	2	2			2	2	2	2	
				2	2	2	2	2			2	2	2	2	2			2	2	2	2	
				2	2	2	2	2			2	2	2	2	2			2	2	2	2	
				2	2	2	2	2			2	2	2	2	2			2	2	2	2	
	< 72			16	16	16	16	64			16	16	16	16	64			16	16	16	16	

Key

W/C = washing and conditioning
< 72 = dwell between conditioning and exposure sequence (Maximum: 72 h)
2 = dump and fill 2 h intervals
16 = 16 h dwell (overnight)
16 = 16 h dwell for data
C = collect prior day's 16 h dwell
64 = 64 h dwell (weekend)

APPENDIX B
EVALUATION OF THE LEACHING POTENTIAL OF ALTERNATIVE BRASSES

Prepared by

Simoni Triantafyllidou and Marc Edwards

Virginia Polytechnic Institute & State University

August 2006

INTRODUCTION

During the Expert Workshop that took place as part of this AwwaRF project, specific research needs were identified. The consensus opinion emphasized the need to *"identify key water quality characteristics in drinking waters that are aggressive to metal release and impact performance in brass"* and *"conduct performance testing with various water qualities and treatments to take into account chloramines and other identified water quality constituents"*.

In an effort to address some of these high priority needs on a preliminary basis, Virginia Polytechnic Institute & State University (VPI) conducted laboratory experiments that examined the leaching potential of seven different alternative brasses. Representative types of Envirobrass® and FederAlloy® alternative alloys, as well as ECO BRASS® were examined (Table B.1).

Table B.1
UNS number and commercial name for the alternative brasses tested in this study.

UNS Number	Trade Name
C89510	Envirobrass® I
C89520	Envirobrass® II
C89837	FederAlloy® I Original
C89831	FederAlloy® I-844
C89844	FederAlloy® I-844 Modified
C89833	FederAlloy® I-836
C69300*	ECO BRASS®

*ECO BRASS® rod. The ECO BRASS® casting alloy, with the same composition as the rod used in this study, has the UNS number C87850

METHODS AND MATERIALS

For each alternative alloy, brass rods of 0.64 cm (or else 0.25 in) diameter and 0.97 cm (or else 0.38 in) height were epoxied to the bottom of a 46 mL glass vial (Figure B.1). This vial was filled all the way to the top with test water and air was excluded. The brass surface area to water volume ratio achieved was 8.0×10^{-3} in^2/mL. This ratio is at the lower end of those typically encountered in NSF testing, assuming ¾ in^2 surface area is typical for a brass faucet (Weed, 2005) and that a typical brass faucet can hold a water volume of 100 mL (Gardels and Sorg 1989). A lower surface area to volume ratio would tend to decrease the final concentration of metals in the water.

All glass vials were kept out of light during the 4-week exposure of alternative alloys to two different leaching solutions. The first leaching solution was the NSF/ANSI Standard 61, Section 9 synthetic test water. This water has alkalinity of 500 mg/L (± 25 mg/L) as $CaCO_3$, pH of 8.0 (±0.5) and free chlorine dosed at a concentration of 2.0 mg/L (± 0.05 mg/L). The second leaching solution, also synthetic, had been identified as aggressive in terms of lead leaching from brass in previous VPI studies. This water has a significantly lower alkalinity of 10 mg/L as $CaCO_3$, a lower pH of 7.4, and chloramines dosed at a concentration of 4.0 mg/L as Cl_2, consistent with changes in U.S. disinfection practice and lower pH and alkalinity water supplies.

The experiment was conducted following a static "dump and fill" protocol. Water was changed inside the glass vials three times a week. At the end of each week water samples were collected for metals analysis. Metals dissolution was quantified using Inductively Coupled Plasma Mass Spectrometry (ICP-MS). In all cases, triplicate samples were tested, to allow for satisfactory Quality Assurance/Quality Control and to give the results statistical power. Since two different water conditions were examined in triplicate for seven alternative brasses, 2 x 3 x 7 = 42 tests were conducted overall, for the non-leaded alloys (Figure B.1).

Figure B.1. Experimental setup, designed to simulate exposure of drinking water to non-leaded brass alloys and assess their leaching potential.

RESULTS

Lead Leaching

Lead leached from all seven alternative brasses was less than 1 ppb throughout the 4-week testing period, and in many instances was below the detection limit (Table B.2). Leaded brass (3% lead) however, which was included in the study for comparison reasons, released much higher lead to the water. In fact, lead dissolution for the conventional alloy was above the Lead and Copper Rule action level of 15 ppb throughout the testing period, with only one exception during week two (Table B.2). For the leaded alloy, the modified water of low pH, low alkalinity and chloramines, was more aggressive towards lead leaching compared to the standard section 9 test water (Table B.2). This observation is consistent with previous tests conducted at Virginia Tech. The leaded brass tested in this study was not treated to remove surface lead, as is sometimes practiced to meet the NSF 61 and California Proposition 65 requirements.

Table B.2
Lead release for the non-leaded brass alloys used in the Virginia Tech study, versus a conventional leaded alloy. The values reported are the average of triplicate samples.

Brass Type	Pb %	Pb Release (ppb)							
		NSF/ANSI 61/ Section 9 Test Water				Low pH, Low Alkalinity, Chloraminated Water			
		Week 1	Week 2	Week 3	Week 4	Week 1	Week 2	Week 3	Week 4
C36000	3.0	23.24	12.63	20.50	21.88	34.96	31.36	37.28	40.46
Envirobrass® I	0.25	0.21	BDL	0.38	BDL	BDL	BDL	0.14	0.11
Envirobrass® II	0.25	BDL	BDL	0.37	0.38	BDL	0.85	0.18	0.72
FederAlloy® I Original	0.1	0.36	BDL	0.27	0.24	BDL	0.85	0.16	0.41
FederAlloy® I 844	0.1	0.76	0.41	0.26	0.31	BDL	BDL	0.07	0.24
FederAlloy® I 844 Mod.	0.2	0.86	0.16	0.47	0.43	BDL	BDL	BDL	0.15
FederAlloy® I 836	0.1	0.60	0.15	0.73	0.34	BDL	BDL	0.05	0.19
ECO BRASS®	0.1	0.36	0.16	0.38	0.39	0.05	0.09	0.27	0.35

BDL stands for Below Detection Limit levels. The detection limit for lead was DL (Pb) = 0.04 ppb

Bismuth Leaching

Bismuth levels were mostly below the detection limit throughout the monitoring period, for both the Section 9 test water, as well as the modified leaching solution (Table B.3).

Table B.3
Bismuth release for the non-leaded brass alloys used in the Virginia Tech study. The values reported are the average of triplicate samples. ECO BRASS® does not contain bismuth, therefore no bismuth data are presented herein for that alloy.

Alternative Brass Type	Bi %	Bi Release (ppb)							
		NSF/ANSI 61/ Section 9 Test Water				Low pH, Low Alkalinity, Chloraminated Water			
		Week 1	Week 2	Week 3	Week 4	Week 1	Week 2	Week 3	Week 4
Envirobrass® I	0.5-1.5	BDL	BDL	0.71	BDL	BDL	BDL	BDL	BDL
Envirobrass® II	1.6-2.2	0.76	BDL	0.97	1.84	BDL	BDL	BDL	BDL
FederAlloy® I Original	0.70-1.2	BDL	BDL	BDL	BDL	BDL	BDL	BDL	BDL
FederAlloy® I 844	2.7-3.7	0.72	BDL	BDL	BDL	BDL	BDL	BDL	BDL
FederAlloy® I 844 Mod.	2.0-4.0	BDL	BDL	BDL	BDL	BDL	BDL	BDL	BDL
FederAlloy® I 836	1.7-2.7	BDL	BDL	BDL	BDL	BDL	BDL	BDL	BDL
ECO BRASS®	-	-	-	-	-	-	-	-	-

BDL stands for Below Detection Limit levels. The detection limit for bismuth was DL (Bi) = 0.70 ppb

Selenium Leaching

Envirobrass® I and II contain selenium. Throughout the study, selenium release from these alloys was below the detection limit of 2.98 ppb for both the Standard 61, Section 9 test water as well as the modified version of that test water.

Copper Leaching

Copper release to the water was below the 1,300 ppb action level for all non-leaded brasses tested, for both exposure to the Section 9 test water, as well as exposure to the modified water (Table B.4).

Table B.4
Copper release for the non-leaded brass alloys used in the Virginia Tech study, versus a conventional leaded alloy. The values reported are the average of triplicate samples.

Brass Type	Cu %	Cu Release (ppb)							
		NSF/ANSI 61/ Section 9 Test Water				Low pH, Low Alkalinity, Chloraminated Water			
		Week 1	Week 2	Week 3	Week 4	Week 1	Week 2	Week 3	Week 4
C36000	60	344.17	321.97	465.90	386.60	41.49	49.88	61.04	75.18
Envirobrass® I	86.0-88.0	638.30	558.93	611.93	725.33	265.60	155.77	177.90	267.03
Envirobrass® II	85.0-87.0	858.37	625.30	638.60	681.03	285.22	157.63	240.17	278.03
FederAlloy® I Original	84.0-88.0	444.64	370.15	429.72	497.24	306.87	197.47	282.80	313.03
FederAlloy® I 844	87.0-91.0	616.20	525.87	664.43	755.80	297.33	162.70	211.90	262.77
FederAlloy® I 844 Mod.	83.0-86.0	595.50	529.20	630.73	741.93	263.23	181.23	155.27	270.57
FederAlloy® I 836	87.0-91.0	595.47	529.77	654.37	729.63	258.57	159.17	198.93	267.07
ECO BRASS®	76	462.03	459.93	469.73	464.83	73.64	129.57	255.60	303.77

Zinc Leaching

Zinc is not currently regulated (no primary or secondary MCL applies for zinc in drinking water) and there is no zinc leachate limit in Standard 61. It is however related to dezincification, a common type of brass corrosion, and as such it was also enumerated in this study. Zinc dissolution for the alternative alloys was in general lower or about the same under exposure to the Standard 61, Section 9 test water, compared to the modified water (Table B.5).

Other Elements

Due to the unusual problem of "carry-over" contamination for antimony via ICP-MS, antimony detection limits were relatively high. When a majority of samples were re-run to minimize this problem, in every sample antimony levels were below the MCL of 6 ppb. Silicon levels leaching from brass were below 500 ppb in all cases.

Visual Differences

Visual differences were observed between the coupons exposed to the Standard 61, Section 9 test water, versus the modified water, for the same alloy. For instance, after completion of the experiment, Envirobrass® II coupons had distinct visual differences between the two waters tested (Figure B.6). In addition, FederAlloy® I Original and ECO BRASS® appeared more uniformly corroded when exposed to the standardized test water (Figures B.7 and B.8).

Table B.5
Zinc release for the non-leaded brass alloys used in the Virginia Tech study, versus a conventional leaded alloy. The values reported are the average of triplicate samples.

Brass Type	Zn %	Zn Release (ppb)							
		NSF/ANSI 61/ Section 9 Test Water				Low pH, Low Alkalinity, Chloraminated Water			
		Week 1	Week 2	Week 3	Week 4	Week 1	Week 2	Week 3	Week 4
C36000	37	579.43	451.70	551.23	679.20	1175.00	576.17	659.27	887.23
Envirobrass® I	4.0-6.0	48.90	86.84	95.73	50.60	53.05	66.62	149.97	110.76
Envirobrass® II	4.0-6.0	50.95	87.30	87.24	56.87	395.03	283.56	210.39	125.52
FederAlloy® I Original	6.0-10.0	39.72	70.52	88.30	65.65	156.33	144.54	138.20	155.70
FederAlloy® I 844	2.0-4.0	27.17	77.90	29.65	40.29	50.46	110.42	111.25	62.76
FederAlloy® I 844 Mod.	7.0-10.0	95.69	48.36	86.36	104.31	89.37	55.60	183.63	115.80
FederAlloy® I 836	2.0-4.0	25.20	49.23	83.68	83.24	42.60	84.45	79.00	49.46
ECO BRASS®	20	451.70	206.07	313.50	292.13	836.63	635.73	574.83	586.40

CONCLUSIONS

The leaching potential of seven non-leaded brasses which are currently commercially available was evaluated in this laboratory study. Lead release from all non-leaded coupons was below 1 ppb throughout the 4 weeks of testing. Bismuth, selenium and antimony were below detection limits. Zinc release into the water tended to depend upon the zinc content of each alloy and copper was below the action limit. Based on these preliminary results, in terms of leaching and contamination concerns, non-leaded alloys provide an attractive alternative for fabrication of drinking water components. Nonetheless, further understanding of their leaching characteristics is needed.

ACKNOWLEDGEMENTS

VPI would like to thank:

- Chase Brass for providing the ECO BRASS® samples
- The Federal Metal Company for providing the FederAlloy® samples
- Colonial Metals for providing the Envirobrass® samples and
- The CDA for providing the 3 percent leaded brass samples

Figure B.6. Untested Envirobrass® II coupon (far left). After exposure to standard Section 9 test water (middle). After exposure to low alkalinity, low pH, chloraminated water (far right).

Figure B.7. Untested FederAlloy® I Original coupon (far left). After exposure to standard Section 9 test water (middle). After exposure to low alkalinity, low pH, chloraminated water (far right).

Figure B.8: Untested ECO BRASS® coupon ((far left). After exposure to standard Section 9 test water (middle). After exposure to low alkalinity, low pH, chloraminated water (far right).

APPENDIX C
EXPERT WORKSHOP SUMMARY

SUMMARY OF EXPERT WORKSHOP

AwwaRF PROJECT 3112 - PERFORMANCE AND METAL RELEASE OF NON-LEADED BRASS METERS, COMPONENTS, AND FITTINGS

DECEMBER 1 AND 2, 2005

NSF INTERNATIONAL, ANN ARBOR, MI

A two day expert workshop was convened for AwwaRF Project #3112. The workshop was convened at NSF International Headquarters in Ann Arbor, Michigan on December 1 and 2, 2005. The purpose of the workshop was to assemble representatives (stakeholders) from a wide variety of groups with interest and expertise in issues surrounding the use of non-leaded brass components in drinking water systems. In-line components used in the transmission and distribution of drinking water include mechanical devices such as valves, meters, backflow prevention devices, pressure regulators, and pump components, and connection devices such as fittings, couplings, meter setters, corporation stops, and curb stops. These stakeholders discussed issues related to non-leaded brass components, (i.e. drinking water regulation, manufacturing, operations and maintenance, cost, performance, testing, and research gaps) and recommended research needs. Following is a summary of discussions and results from this workshop. Results from this workshop will be used to identify and prioritize research needs related to non-leaded brass components. In addition, a preliminary structure will be developed for the top-priority recommended research projects.

LIST OF ATTENDEES

Thirty eight people attended the workshop. Table C.1 contains a list of attendees and their affiliations

Table C.1
List of Workshop Participants

Participant	Affiliation
Rich Brown	Portland Water Bureau
Bob Powell	Pinellas County Water
Jack DeMarco	Greater Cincinnati Water Works
Michael Schock, PhD	USEPA
Lisa Donahue	USEPA Region 3 (3WP32)
France Lemieux	Health Canada
Lance Agness	Ford Meter Box
Mike Jones	Ford Meter Box
Craig Selover	Masco Corporation
Lee Mercer	Moen Incorporated
Brad Homiston	Kohler Company
Todd Kidder	Federal Metal Company

(Continued)

Table C.1 (Continued)
List of Workshop Participants

Participant	Affiliation
Joseph Napolitan	Extruded Metals Company
Barry Munce	Extruded Metals Company
Harry H. Libengood, Jr.	Concast Metal Products Co.
Dom Lemaire	Concast Metal Products Co.
Richard J. Emerick	Elkay Manufacturing Company
Devin K. Denner	Chase Brass
Larry Muller	Chase Brass
Myron Lee	Mueller Brass
Mike Buyarski	Colonial Metals
Tim Bianchi	Neptune Meter
Don Kullman	Neptune Meter
Pete Greiner	NSF
Clif McLellan	NSF
Jeremy Brown	NSF
Bob Weed	Copper Development Association
Andy Kireta	Copper Development Association
Dave Viola	Plumbing Manufacturers Institute
Marc Edwards, PhD	Virginia Tech
Simoni Triantafyllidou	Virginia Tech
Gregory Kirmeyer	HDR
Anne Sandvig	HDR
Glen Boyd	HDR Engineering, Inc.
Dan Laprade (PAC)	Mass. DEP
David Heumann (PAC)	Los Angeles Dept. of Water & Power
Jeff Kempic (PAC)	USEPA
Traci Case	AwwaRF

AGENDA

A workshop package was send to each attendee two weeks prior to the workshop. This package contained an agenda, background materials, and a series of draft topic papers prepared by HDR/EES and Virginia Polytechnic Institute and State University (VT). These topic papers were organized into the following subtopics:

- Description of the Composition, Lead Content and Mechanical Properties of New Non-Leaded Brasses
- Potential of Non-Leaded Brasses to Release Lead, Copper, and other Metals (Bismuth and Selenium) into Drinking Water
- Potential Health Effects of Non-Leaded Brasses
- ASTM Specifications for New Non-Leaded Brasses

- Summary of Current State Experiences with Non-Leaded Materials
- Utility Motivations and Experience with Non-Leaded Brass Components
- Potential Impacts to Manufacturers
- Plumbing Industry Issues
- Potential Impacts to Utility Operations and Maintenance Procedures
- Inventory and Drawbacks of Other Available Materials Not Currently Considered for Fabrication of Fixtures and Components

Following introductory comments by the workshop organizers, sponsors and hosts, several presentations were made by individuals who summarized the information contained in the topic papers. Small group breakout sessions were conducted during the two-day workshop to identify data needs and formulate specific research objectives. After the small-group breakout sessions, the entire group re-convened for presentations from each break-out group and general discussion.

DAY 1 - WELCOME, INTRODUCTIONS, AND WORKSHOP OBJECTIVES

Anne Sandvig introduced Project Manager Traci Case, AwwaRF and Cliff McClellan, NSF International. Traci welcomed the group and identified the AwwaRF Project Advisory Committee (PAC) members: Dan Laprade, David Heumann, and Jeff Kempic. Cliff McClellan of NSF International also welcomed the group and provided some background on the certification role of NSF International.

ISSUE SUMMARIES/PRESENTATIONS

Anne Sandvig introduced several speakers who provided summary presentations of the key issues and information contained in the draft topic papers. The first speaker was Simoni Triantafillydou, VPI who presented "*Composition, Leaching, Alternative Materials/Processes*". Comments from participants included correcting the free machining ranges (Craig Selover) and discussions of the Cincinnati study on leaching from curb stops. Bob Weed, CDA, presented "*Potential Impacts to Manufacturers*", which was followed by a questions/comment period. Lee Mercer asked what the USGS assumption was on the availability of bismuth, and Bob responded that the USGS assumptions did not evaluate the scenario of bismuth replacing all lead in alloys. Craig Selover asked what survey information on leaded brass in the plumbing industry could be used to evaluate supply/demand capacity, and Dave Heumann, LADWP also asked about supply/demand capacity. Bob stated that USGS says there is an adequate supply worldwide. Lance Agness discussed an AFS study which contains an estimate on bismuth supply using Bolivian (low grade) versus Chinese mines. Mike Buyarski brought up other available products and asked for opinions on bismuth recycling. Bob stated the need to separate lead versus bismuth alloys during recycling.

Dave Viola presented "*Impacts and Issues of Moving the Plumbing Industry Towards the Use of Non-Leaded Brass*". Marc Edwards asked how cheaters (i.e. installers that continue to install non-certified components) were handled and Dave answered that each state/city is different but he has experienced enforcement activities where illegal suppliers have been shut down. The concern is where and how is enforcement handled. There is a need for public education and third party certification. France Lemieux discussed what PMI's and certification agencies roles were in dealing with counterfeit components. Pete Greiner said that manufacturers who mislead purchasers by stating their products are certified by a testing laboratory and place the testing lab certification mark on their product (or product literature) when really they are not. NSF, and likely all certifying agencies, aggressively pursue illegal uses of the mark like this. NSF re-tests in-line devices and adoption of model plumbing codes changes or manufacturing retooling/redesigning or products is typically a $\pm$ 3 year process. Lisa Donahue discussed traceability for identifying counterfeit products; post installation with certified in-line devices versus certified end-use devices and how to trace plumber (customer) vs utility installed components. Dave Viola suggested use of the NSF web page for identifying certified products.

Anne Sandvig presented "*Regulatory Issues*", a summary of federal and state regulations pertaining to materials use in utility distribution systems, followed by Dave Heumann who presented background information on California Proposition 65. Lee Mercer asked if selenium and/or bismuth were on the Proposition 65 list and Dave answered that he did not know, however zinc was on this list.

After a short break, Anne Sandvig presented "*Utility Experiences with Non-leaded Brass*". Dave Heumann emphasized the importance of putting cost information into the final report, currently the best information is from Cincinnati (cold climate) and LADWP (warm climate). Dan Laprade noted that, for many utilities, the switch to non-leaded components would be quicker if there were data that showed an improvement in the utility's 90th percentile after making the change to nonleaded components. Utilities, especially small and medium sized (<50,000 population) are not likely to make the change if it is going to cost them more and not have any net effect on their 90th percentile lead level. Jack DeMarco agreed. Rich Brown discussed the lengthy (1 year) process involved in getting budget approvals from the city council and Jack Demarco brought up the time needed to conduct in-house testing. Anne Sandvig suggested that the time frame for code revisions and utility input should be taken into account when determining time frames for implementing non-leaded replacement programs at utilities. Dave Heumann felt that the more people buy non-leaded components, the more demand for these products. Since meters are replaced periodically, it would be useful for utilities to start their replacement programs now. However, there is a need for more data on how often the slug of lead might be contained in the one liter regulatory sample collected at the tap. Tim Bianchi said that there are only three water meter manufacturers in North America and that Neptune sells on third of all water meters. Since 2001 they have exclusively sold non-leaded meters. California Proposition 65 forced the change because Neptune didn't want to use two alloys in manufacturing, i.e. they wanted to use just one foundry. Rich Brown said he was told by sales representatives that some utilities don't want to use the non-leaded components and also that the use of non-leaded components becomes a scrap metal issue on the recycling side.

The presentations were followed by an open discussion session from 11:15am. until noon. Anne Sandvig summarized the major issues identified in the topic papers and presented in the issue summary presentations. These issues were categorized into the following major groups: Manufacturing, Plumbing Component Industry, Regulatory, and Utility. Discussion followed. For the Manufacturing Issues, Dave Viola discussed different approaches for reducing lead leaching from components, i.e. performance testing versus specification of products with lower lead content. Rich Brown brought up the coating of alloys with respect to alleviating the leaching problem. Craig Selover wanted AWWA performance versus NSF/ANSI standards to not be considered as isolated regulations/standards and looked at as a whole. Larry Muller asked about chemical treatment versus lead source replacement and Anne provided some background on steps utilities have been required to take over the last several years to reduce lead in drinking water. Larry Muller presented information on the silicon brass his company produces, which contains a maximum lead content of 0.1 percent lead. In Japan, 11 million pounds have been produced since 2000, and 5 percent of all water meters sold are made using this alloy. Brad Homiston suggested the need for independent third party verification of machinability to compare alloys head-to-head and the need to define machinability and methods of testing for machinability.

For the Materials Testing and Certification Issues, Dave Viola discussed the availability of several laboratories for obtaining certification of products. Mike Buyarski asked about health effects of selenium and bismuth and Anne mentioned that that specific issue would be handled in the Regulatory break-out group. Dave Viola mentioned that product standards are already referenced in codes but there needs to be a strategy for including health based performance standards in project standards. Pete Greiner reminded the group that Standard 61 is a risk assessment based certification process and the standard is not just a lead standard, but rather it evaluates any extractants, including bismuth and selenium. Rich Emerick emphasized the need to discuss what information may be needed on the change in component leaching characteristics with future changes in water quality (i.e. chloramines).

Discussion on Regulatory Issues included understanding the need to re-write regulations in some cases (Lisa Donahue) as well as the relationship between utility specification procedures and the local plumbing authorities (Craig Selover). Andy Kireta stressed the importance of the impact of other regulations on non-leaded brass (simultaneous compliance issues). Marc Edwards discussed San Francisco and LADWP converting to chloramines and that stress corrosion can be affected by changes in water quality, i.e. change in disinfectants. Dan Laprade mentioned the need for consistency between the various state regulations, and Mike Schock discussed the importance of understanding the feasibility and effectiveness of various regulatory approaches.

For the Utility Issues, Richard Emerick emphasized the need to educate smaller utilities about the reasons for buying non-leaded components. Mike Buyarski asked for data showing the effectiveness of using non-leaded components and Rich Brown asked what the goal for effectiveness should be. Bob Powell asked if there should be a recommendation in an AWWA topic paper that states "use non-leaded" materials.

With respect to Plumbing Component Industry Impacts, Craig Selover and Bob Powell mentioned the importance of enforcement and the need for inspectors and potentially an accounting review to determine if specified items were actually purchased.

FIRST BREAKOUT GROUP SESSION – IDENTIFY ISSUES

After the general discussion, the group was divided in five break-out groups based on which sector of the waterworks industry each participant represented. Table C.2 contains a list of each break-out group. The objectives for the breakout groups were to:

1. Identify the potential issues presented by non-leaded brass to your sector of the water industry (using initial issues/questions prepared by project team as a starting point), and
2. Prioritize these issues to the extent possible (top 5).

An initial list of issues/questions were prepared for each group to serve as a starting point for meeting the objectives. At 2:45 p.m. the groups reconvened and each group presented their prioritized issues.

Table C.2
Breakout Groups

Breakout Group	**Name**	**Affiliation**
Utility Group Gregg Kirmeyer, Moderator	Rich Brown	Portland Water Bureau
	Bob Powell	Pinellas County Water
	Jack DeMarco	Cincinnati Water Works
	Dave Heumann	LADWP (PAC)
	Traci Case	AwwaRF
Regulatory Group Marc Edwards, Moderator	Michael Schock	USEPA – ORD
	Jeff Kempic	USEPA (PAC)
	Lisa Donahue	USEPA Region 3
	France Lemieux	Health Canada
	Dan Laprade	MADEP (PAC)
	Simoni Triantafillydou	Virginia Tech
	Cliff McLellan	NSF International
Manufacturing Group Bob Weed, Moderator	Todd Kidder	Federal Metal
	Davis Anderson	Chicago Extruded Metals
	Joseph Napolitan/Barry Munce	Extruded Metals Company
	Harry H. Libengood Jr./Don Lemaire	Concast Company
	Devin K. Denner/Larry Muller	Chase Brass
	James Rourke/Myron Lee	Mueller Brass
	Mike Buyarski	Colonial Metals
Plumbing Component Industry Group Anne Sandvig, Moderator	Lance Agness	Ford Meter Box
	Mike Jones	Ford Meter Box
	Craig Selover	Masco Corporation
	Lee Mercer	Moen
	Brad Homiston	Kohler
	Tim Bianchi/Don Kullman	Neptune Meter
	Richard Emerick	Elkay Manufacturing
Testing/Professional Organization Group Glen Boyd, Moderator	Pete Greiner	NSF International
	Jeremy Brown	NSF International
	Andy Kireta	Copper Development Association
	Dave Viola	Plumbing Manufacturer's Institute

Utility Group Session I Report

Bob Powell presented the priority issues as identified by the utility group. They were (in order of priority):

1. *Performance of non-leaded brass components.* Both short-term and long-term evaluations of performance should be considered. Short-term includes approval testing and long-term should include long term monitoring, structural, mechanical, operability, tracking, and accelerated testing
2. *Costs.* Costs of implementing non-leaded brass component replacement programs should include unit costs, and a sensitivity analysis associated with failure of components already in the ground.

3. *Incentives/Encouragement.* Should quantify health benefit of low lead-leaching components for consumers. Education of various stakeholders: utility, decision makers/regulators, code enforcers at all levels. Tools to use could include AWWA policy, USEPA, USCDC, NIH, ASTDR, ASDWA

Discussion comments/questions included Dan Laprade stating that many small systems are putting in meters for the first time and that many small and medium utilities will simply emulate what the large utilities are doing (with respect to using non-leaded components).

Regulatory Group Session I Report

Marc Edwards presented prioritized issues from the regulatory group, which were:

1. *Health Concerns.* For all alternatives to leaded brass, do the new potential leachates (P, Si, Bi, Se) pose health concerns? Include analysis of occurrence, likely future regulation, radioactivity, oxidation states. Compare back to lead, with existing data and knowledge.
2. *Feasibility/Effectiveness of various Regulatory Approaches.* Safe Drinking Water Act (e.g., percent lead specification and health based performance standards), State Regulations and programs, and/or plumbing code. Consider extension of health based leaching performance standards to all components. Regulate at point of sale (buyer/seller) as per pesticides? Achieve consistency in approach while allowing innovation.
3. *Plumbing Code Revisions.* Should plumbing codes such as the IPC and the UPC be revised to address the lead content of drinking water components or should health based performance standards be extended to cover all products? If so, what steps would need to be taken?
4. *Regulatory Timelines and Approach.* Timelines and approach for changing and implementing new regulations related to non-leaded brass should be considered.
5. *Impact of Regulations.* Consider impacts of current and new regulations on performance of plumbing components with very low lead-leaching potential (health and mechanical). Consider changes in ammonia and mineral composition.
6. *Assisting State Regulatory Programs.* What can be done at the federal level to assist states in developing and implementing programs related to use of non-leaded brass components in drinking water systems?
7. *Non-regulatory Approaches.* Can non-regulatory approaches influence utility purchasing decisions to achieve goals of installing components with very low lead-leaching potential
8. *Enforcement.* Enforcement of Non-Permitted Rehab (home remodeler)

Mike Schock emphasized the need for a three-pronged regulatory approach which addresses inconsistencies between federal and state regulations, updating plumbing codes, and monitoring purchase/transactions at the consumer level. He also mentioned the differences in Section 9 versus Section 8 of NSF/ANSI Standard 61.

Manufacturing Group Session I Report

Larry Muller presented the prioritized issues from the manufacturing group. They are summarized below.

1. *Impurities in Recycled Materials.* There are incidental impurities in recycled materials – Lead vs Alternative alloys. This is consistently a problem and can have a dramatic effect on customer relations.
2. *Quality of Imported Products.* There are quality concerns with imported products (unfair competition). A level playing field and global enforcement are required.
3. *Bismuth Supply.* Understanding the near-term and long-term supply of bismuth
4. *Manufacturing Limitations of Alternative Materials.* Raw material, process and/or manufacturing limitations of alternative materials
5. *Machinability Benchmarking.* Machinability is important in marketing to users and to product acceptance in the marketplace
6. *Understanding Cast vs. Wrought Products.* Lack of understanding of the difference between cast and wrought products
7. *Standards.* Performance standards as opposed to prescriptive standards. Consistency of performance standards would facilitate manufacturing consistency.

Rich Brown asked what percentage of all brass manufactured is used in the drinking water industry, and the answer was 35 – 45 percent. Gregg Kirmeyer asked about machinability benchmarking. Apparently, there is an ASTM standard for machinability of ferrous products (E618 ASTM) but nothing for brass. There are machinability numbers available, but no-one was sure where those values came from and there was a general consensus that the industry needed benchmarks and consistency in determining machinability. Bob Weed mentioned that the CDA has modified machinability standards for rod products. With respect to standards, Andy Kireta stated that there was a process in place (NSF/ANSI Standard 61) that included health effects evaluations, third party certification and validation and that no change to standards was needed. Both Craig Selover and Andy Kireta stated that the standard is dynamic and responsive, and that these standards are flexible. Gregg Kirmeyer asked about validation/field verification and who determines health level of contaminant. There was a general discussion about the NSF criteria and requirements for health risk (toxicological criteria) and field verification of processes and products by third party certifiers. NSF has a health advisory board with external peer review by toxicologists.

Plumbing Component Industry Group Session I Report

Craig Selover presented the prioritized issues determined by the Plumbing Component Industry Group. They were as follows:

1. Availability of Se/Bi Alloys and Copper. Virgin vs. scrap
2. Health Effects of Bismuth.
3. Performance based acceptance criteria. Including NSF/ANSI Standard 61 and performance based standards.
4. Processibility of Available Alloys. With respect to both cost and product performance.
5. Long-term Performance. Both external and internal corrosion.

Testing/Professional Organization Group Session I Report

Andy Kireta presented the prioritized issues identified by the Testing/Professional Organization Group. He emphasized that there is a certification process in place and that no changes in standards were needed, however the following major issues were identified:

1. Health effects (NSF/ANSI Standard 61) for all products
2. Third party certification
3. Validation
4. Evaluating changes in water chemistry and impacts on non-leaded brass components

Craig Selover mentioned that the standard is dynamic and responsive and needs to change if necessary. Andy reiterated that the standards are in place to be flexible. Gregg Kirmeyer asked for a definition of "validation" and who determines the health level of a contaminant for purposes of NSF certification. Andy and Pete Greiner responded that "validation" meant field verification of NSF certification results, and that NSF has a health advisory board along with an external peer review panel of toxicologists to determine health levels.

DAY 2 – SECOND BREAKOUT GROUP SESSION – RESEARCH NEEDS

The group was again divided into the same five break-out groups for the second day of the workshop. The objectives for this second breakout group session were to identify research needs for the prioritized list of issues identified in Session I. Research Prioritization Worksheets were handed out to each group to assist them in identifying the details of each research need. Each group was asked to complete worksheets for at least two of the identified research needs. At 10:25 a.m. the groups reconvened and each group presented their research needs.

Manufacturing Group Session II Report

The data needs identified by the manufacturing group are described below.

Recycling

- Understanding the effect of bismuth is the key concern
- What is the size of the non-leaded cast alloy market?
- Cost of sending scrap all the way back to refiners
- What is the potential that product might be sent to a landfill instead of recycled (we think this is a bad idea!)
- Investigate how other industries handle dissimilar scrap streams
- Potential additional contacts: ISRI, bismuth refiners.

Global Materials Supply

- Global Supply-Demand study (End Product), include Bi, Se, Si, Sb
- Is bismuth recyclable?
- What are the major forms of bismuth scrap?

Machinability Standards

- Responsibility should reside within the copper industry
- Develop a useful machinability model
- Coordinate standard development with PMI members and other material users (machining processes)

Point of Sale Auditing

- Tracing point-of-sale paper-work followed by testing actual faucet

There was general discussion about whether the manufacturing community can give the market what they want, and are they giving the market what it wants. Mike Schock emphasized that the point-of-sale auditing issue needs to be integrated with regulatory issues. This was followed by discussion on enforcement authority and accreditation. Lisa Donahue discussed possible similarities with the FIFRA program for pesticides, where there is legal authority to enforce sale of pesticides, but the SDWA does not provide a legal mechanism for controlling drinking water components. Bob Weed emphasized that the highest priority research need from this group was the recycling issue.

Plumbing Component Industry Group Session II Report

Craig Selover presented the research needs identified by the Plumbing Component Industry Group, which were as follows:

Availability of Se/Bi Alloys and Copper
Health Effects of Bismuth
Performance Based Acceptance Criteria

Utility Group Session II Report

Gregg Kirmeyer presented the research needs from the utility group, as follows:

Short-term Performance (months – 1 ½ yr)

- Utilizing the laboratory at NSF
- Evaluating leachability of non-leaded components (Se, Bi, etc.)
- Evaluation of Coatings
- Field evaluation of in-place components, i.e. gather information from utilities that have installed these materials such as LADWP
- Evaluate structural/mechanical performance
- Identify new testing methods

Long-term performance (20 yrs)

- Develop a clearing house of information contained at central location (AwwaRF, AWWA)
- Conduct testing at large utilities (such as leaching or failure of components)

Costs (Narrow)
- Unit costs for all in-line devices
- Unit costs for in-line non-leaded components
- Develop combined costs for all materials
- Combine labor/materials costs (30 percent/70 percent)
- Life Cycle Costs
- Long-term scheduling

Cost (Broad)
- Global impact analysis (e.g. recycling)
- Recycle impacts
- Manufacturing constraints

Incentive/Encouragement
- Data gathering on health impacts by USCDC, such as correlation of lead in water to blood lead levels.
- Develop educational tools, such as technology transfers, using trade organizations such as AWWA, to encourage cross-pollination of information with the public, i.e. decision makers

Rich Brown from Portland suggested that long-term data collection include mileage on water meters. Pete Greiner discussed Standard 61 and the fact that the risk assessment is already included, for all extractants, not just lead.

Regulatory Group Session II Report

Marc Edwards presented three research needs topics from the regulatory group, which included;

Feasibility/Effectiveness of Various Regulatory Approaches
Impacts of changes in raw and treated water quality on performance(health and mechanical) of plumbing components with very low lead-leaching potential
Characterization and relative toxicological comparison of lead substitutes in brass (Si,Bi, P, Se, etc).

Testing/Professional Organization Group Session II Report

Jeremy Brown presented the research needs from this group, which included research on:

Aggressiveness of real waters
Leaching of non-leaded alloys in real waters
Long term performance of non-leaded alloys

Jeremy initially discussed general materials testing, certification and standards and that protocols and criteria are currently in the standards and are adequate, i.e. special protocols are not necessary. However, there is a need to require NSF/ANSI Standard 61 compliance for all products, 3rd party certification, performance testing with various water qualities and treatment (non-health issues) to take into account chloramines and other identified water quality constituents, and validation of NSF/ANSI Standard 61.

There was a discussion after the presentation about understanding corrosion of drinking waters for a range of supplies and how the quality of private wells might fit into the overall picture.

MEETING WRAP-UP/ADJOURNMENT

Participants were asked to provide comments by December 16, 2005 on the Draft Topic Papers contained in the Workshop Packet. The project team will provide a summary/minutes of the workshop and send that out to all participants. The meeting was adjourned at approximately 12:00 p.m. on December 2, 2005.

APPENDIX D
SUMMARY OF PRIORITIZED RESEARCH ISSUES

Table D.1
Summary of Prioritized Research Issues Identified in Expert Workshop

Major Subject Area	Research Need	Break-out Group that Identified the Research Need	Priority Identified by Break-out Group (1 is the highest)
LEACHING, PERFORMANCE, AND PRODUCT ACCEPTANCE CRITERIA	*Performance of Non-leaded Brass Components.* Both short-term and long-term evaluations of performance should be considered. Short-term includes approval testing and long-term should include long- term monitoring, structural, mechanical, operability, tracking, and accelerated testing.	Utility Group	1
	Changes in Water Chemistry and Metal Release for all Products Identify key water quality characteristics in drinking waters that are aggressive to metal release and impact performance in brass. Conduct performance testing with various water qualities and treatments to take into account chloramines and other identified water quality constituents.	Testing/Professional Group	1 and 2
	Impacts of changes in raw and treated water quality on performance (health and mechanical) of plumbing components with very low lead-leaching potential	Regulatory	1
	Obtain data on Impacts of aggressive water qualities on long-term performance of products using accelerated methods	Testing/Professional Group	5
	Long-term Performance. Performance of products related to both external and internal corrosion.	Plumbing Group	5
	Standards Development of consistent standards. Consistency of standards would facilitate manufacturing	Manufacturing Group	7
COSTS	*Costs – Narrow* Costs of implementing non-leaded brass component replacement programs - including unit costs and sensitivity analyses associated with failure of components already in the ground.	Utility Group	2
	Costs – Broad Broader economic analyses related to recycling, supply, and manufacturing limitations on non-leaded brass alloys and components.	Utility Group and Manufacturing Group	2

(continued)

Table D.1 (Continued)
Summary of Prioritized Research Issues Identified in Expert Workshop

Major Subject Area	Research Need	Break-out Group that Identified the Research Need	Priority Identified by Break-out Group (1 is the highest)
PLUMBING CODE REVISIONS AND MARKETPLACE ENFORCEMENT	*Plumbing Code Revisions.* Should plumbing codes such as the IPC and the UPC be revised to address the lead content of drinking water components or should health based performance standards be extended to cover all products	Regulatory Group	2
	Quality of Imported Products. There are quality concerns with imported products (unfair competition). A level playing field and global enforcement are required.	Manufacturing Group	2
	Third Party Evaluation and Validation of NSF/ANSI Standard 61 Evaluation of the status of validation and third party certification of NSF/ANSI 61 standards	Testing/Professional Group	3 and 4
	Marektplace Enforcement. Enforcement of Non-Permitted Rehabilitation (home remodelers)	Regulatory Group	7
EDUCATION AND TECHNOLOGY TRANSFER	*Education/Incentives.* Quantify and communicate the health benefit of low lead-leaching components for consumers. Education of various stakeholders: utility, decision makers/regulators, code enforcers at all levels	Utility Group	3
	Performance based acceptance criteria. Performance Based Acceptance Criteria – Generating information and providing educational opportunities to express the importance of performance criteria to waterworks industry.	Plumbing Group	3
	Understanding Cast vs. Wrought Products, etc. Education and public information to correct the lack of understanding of the difference between cast and wrought products, fixtures vs. fittings, etc. and alloys currently available for use.	Manufacturing Group	6

(Continued)

Table D.1 (Continued)
Summary of Prioritized Research Issues Identified in Expert Workshop

Major Subject Area	Research Need	Break-out Group that Identified the Research Need	Priority Identified by Break-out Group (1 is the highest)
MANUFACTURING ISSUES	*Impurities in Recycled Materials.* Understanding how impurities in the recycling stream caused by the increased use of non-leaded materials may impact the manufacture of alloys and components	Manufacturing Group	1
	Availability of Se/Bi Alloys and Copper Impacts of the global supply of selenium and bismuth and copper on the increased demand for non-leaded components, including virgin raw material as well as recycled scrap material	Plumbing Group	1
	Bismuth Supply. Understanding the near-term and long-term supply of bismuth and the supply/demand, on a global basis, of alloying elements used in non-leaded brass	Manufacturing Group	3
	Manufacturing Limitations of Alternative Materials. Raw material, process and/or manufacturing limitations of alternative materials	Manufacturing Group	4
	Processibility of Available Alloys. What are the constraints related to processing and manufacturing components made from non-leaded alloys. What are the cost implications related to processing these alloys and how would the end products perform.	Plumbing Group	4
	Machinability Standards/Benchmarking Develop a universal machinability standard for non-leaded brass alloys.	Manufacturing Group	5

(Continued)

Table D.1 (Continued)
Summary of Prioritized Research Issues Identified in Expert Workshop

Major Subject Area	Research Need	Break-out Group that Identified the Research Need	Priority Identified by Break-out Group (1 is the highest)
REGULATORY IMPACT AND IMPLEMENTATION	*Health Concerns.* For all alternatives to leaded brass, do the new potential leachates (P, Si, Bi, Se) pose health concerns? Include analysis of occurrence, radioactivity, oxidation states. Compare back to lead, with existing data and knowledge.	Regulatory Group	1
	Health Effects of Bismuth. Understanding the health effects of bismuth and how these impacts compare to lead.	Plumbing Group	2
	Regulatory Timelines and Approach. Timelines and approach for changing and implementing new regulations related to non-leaded brass should be considered.	Regulatory Group	3
	Impact of Regulations. Consider impacts of current and new regulations on performance of plumbing components with very low lead-leaching potential (health and mechanical).	Regulatory Group	4
	Assisting State Regulatory Programs. What can be done at the federal level to assist states in developing and implementing programs related to use of non-leaded brass components in drinking water systems?	Regulatory Group	5
	Non-regulatory Approaches. Can non-regulatory approaches influence utility purchasing decisions to achieve goals of installing components with very low lead-leaching potential	Regulatory Group	6

REFERENCES

Agency for Toxic Substances and Disease Registry. 1995. *ToxFAQs for Antimony*. Atlanta, GA: U.S. Department of Health and Human Services, Public Health Service.

Agency for Toxic Substances and Disease Registry. 2003a. *Toxicological Profile for Selenium*. Atlanta, GA: U.S. Department of Health and Human Services, Public Health Service.

Agency for Toxic Substances and Disease Registry. 2003b. *Public Health Statement for Selenium*. Atlanta, GA: U.S. Department of Health and Human Services, Public Health Service.

Agency for Toxic Substances and Disease Registry. 2003c. *Public Health Statement for Zinc*. Atlanta, GA: U.S. Department of Health and Human Services, Public Health Service.

Agency for Toxic Substances and Disease Registry. 2003d. *ToxFAQs for Zinc*. Atlanta, GA: U.S. Department of Health and Human Services, Public Health Service.

Agency for Toxic Substances and Disease Registry. 2003e. *ToxFAQs for Nickel*. Atlanta, GA: U.S. Department of Health and Human Services, Public Health Service. http://www.atsdr.cdc.gov/tfacts15.html. Accessed 11-9-2005.

Agency for Toxic Substances and Disease Registry. 2004. *Public Health Statement for Copper*. Atlanta, GA: U.S. Department of Health and Human Services, Public Health Service.

Agency for Toxic Substances and Disease Registry. 2005a. *Public Health Statement for Tin*. Atlanta, GA: U.S. Department of Health and Human Services, Public Health Service.

Agency for Toxic Substances and Disease Registry. 2005b. *Public Health Statement for Lead*. Atlanta, GA: U.S. Department of Health and Human Services, Public Health Service.

Anderson G. 1994. "One Manufacturer's Perspective on Complying with the NSF/ANSI Standard 61 and The Lead Leaching Issue". *Proceedings American Water Works Association Annual Conference*, New York, June 19-23, 1994.

Anonymous, 1951. "Review of Current Investigation No 8: Waterworks Fittings. Research Group Report." *J. Inst. Water Engrs*. (Br.), 5:700.

Asarco. 2000. *Asarco Water Utilities Survey Report*. Asarco Globe Plant. Denver, Co. .

ASTM. 1994. ASTM G37-90, Standard Practice for Use of Mattsson's Solution of pH 7.2 to Evaluate the Stress Corrosion Cracking Susceptibility of Copper-Zinc Alloys 129-131.

ASTM. 2005. Standard A 380-99. 2005 (Re-approved). *Standard Practice for Cleaning, Descaling, and Passivation of Stainless Steel Parts, Equipment, and Systems*. ASTM International. West Conshohocken, PA

ASTM International. 2005. "ASTM International Overview." Accessed at http://www.astm.org/cgi-bin/SoftCart.exe/ABOUT/aboutASTM.html?L+mystore+fswh5103+1131695471 on 11/10/2005.

Bangor Water District. 2005. "Material Specifications." Accessed at: http://www.bangorwater.org/pdf%20files/Construction%20Material%20Specifications,%20Effective%20%203-19-04.pdf

Bellinger, D.C. 2003. "Lead". *Pediatrics*. Vol. 113, Issue 4, page 1016. Evanstan, IL.

Boyd, G.R.., G.J. Kirmeyer, G.L. Pierson, S.L. Hendrickson, D. Kreider, and R. English. 2005. "Testing of Point-of-Use Filters at Seattle Schools Drinking Fountains", presented at *Awwa Water Quality Technology Conference*, Quebec City, Canada, Nov. 6-10.

Canfield, R,, C. Henderson, D. Cory-Slechta, C. Cox, J. Todd, and B. Lanphear. 2003. "Intellectual Impairment in Children with Blood Lead Concentrations below 10 ug per Deciliter". *N. Engl. J. Med.* 348:1517-26.

Cassidy, V. 1994. "Industry gets the lead out." *Modern Metals*, Vol. 50, Issue 9, October 1994.

Chase Brass and Copper Company, Inc. 2006. Letter to Glen Boyd (HDR) dated January 11, 2006.

Copper Development Association (CDA). 1996. "Lead-free Cast Red Brasses for Plumbing. Copper Topics, Spring 1996." www.copper.org/resouces/cutopics/Ct81/castred.html .

Copper Development Association (CDA). 2000. *Alloy Data Sheet, EnviroBrass® (SeBiLOY®), Non-Leaded Red Brass and Yellow Brass Casting Alloys, Publication Number A1032-95/00.* The Copper Development Association Inc., New York, 2000. Accessed at http://www.copper.org/applications/industrial/PDF_files/sebiloy.PDF on 9/25/2005

Copper Development Association (CDA). 2005. *Mechanical Properties of Wrought and Cast Copper Alloys, Copper Development Association, 2005.* Accessed from the Properties of Wrought and Cast Copper Alloys Search at http://www.copper.org/resources/properties/db/CDAPropertiesSelectionServlet.jsp?mode=basic on 10/28/2005.

De Marco, J. 2005. "Utility Challenges". Presented at *US EPA Lead and Copper Rule Workshop: Lead in Plumbing Fittings and Fixtures*, July 26-27, 2005.

Dresher, W.H. and D.T. Peters. 1992. "Lead-Free and Reduced Lead Copper-Base Plumbing Alloys: Evaluation of Candidate U.S. Compositions." Unpublished Report, pp 37. International Copper Association (ICA). New York, New York.

Federal Metal Company. 2005. "Composition of FederAlloy® I, II, II, IV and V." The Federal Metal Company. Accessed at http://www.federalmetal.com/fedalloy/alloy1.htm, http://www.federalmetal.com/fedalloy/alloy2.htm, http://www.federalmetal.com/fedalloy/alloy3.htm and http://www.federalmetal.com/fedalloy/alloy4.htm on 9/25/2005

Ford Meter Box. 2005a. *A Manufacturer's Perspective on a Reduction In the SDWA Definition of Lead-free to 0.2 Percent.* Wabash, IN: Ford Meter Box, Inc.

Ford Meter Box. 2005b. *Position on Proposed Rulemaking Re: Lead in Brass Waterworks Fittings.* Wabash, IN: Ford Meter Box Inc.

Ford Meter Box Company, Inc. 2005c. "The Kunkel Process Validation". September 23, 2005. Accessed at http://www.fordmeterbos.com/catpdfs/Kunkel_Process_Validation_Paper_2005.pdf on 10/20/2005.

Gardels, M.C. and T.J. Sorg. 1989. "A Laboratory Study of the Leaching of Lead from Water Faucets". *Jour. AWWA*, 81 :7 :101 (1989) .

George, M.W. 2003a. "Selenium and Tellurium." *U.S. Geological Survey Minerals Yearbook.* U.S. Geological Survey, Reston, VA.

George, M.W. 2003b. "Bismuth." *U.S. Geological Survey Minerals Yearbook.* U.S. Geological Survey, Reston, VA.

Gilbert, S.G. and B. Weiss. 2006. "A Rationale for Lowering the Blood Lead Action Level from 10 to 2 μg/dL." *NeuroToxicology* 27:693-701.

Giordano, C.M., G.S. Duffo, and J.R. Galvele, 1997. "The Effect of Cu+2 Concentration on the Stress Corrosion Cracking Susceptibility of ∝-Brass in Cupric Nitrate Solutions." *Corrosion Science.* V. 39, N. 10-11 1915-1923.

Greiner, Pete. 2006. Unpublished report prepared for NSF International. Ann Arbor, Michigan. August, 2006

Guo, X., W. Chu, K. Gao and L. Qiao, 2002. "Relativity between corrosion-induced stress and stress corrosion cracking of brass in an ammonia solution." *Journal of University of Science and Technology.* Beijing, 9:6:431.

HDR/EES 2005a. Unpublished survey results for Awwa Research Foundation project #3018, "Contribution of Service Line and Plumbing Fixtures to Lead and Copper Rule Compliance Issues". HDR Engineering, Bellevue, WA.

HDR/EES 2005b. LADWP Lead Source Replacement Case Study. Draft Report for AwwaRF Project No. 3018. HDR Engineering, Bellevue, WA.

HDR/EES. 2005c. EBMUD Lead Source Replacement Case Study. Draft Report for AwwaRF Project No. 3018. HDR Engineering, Bellevue, WA.

HDR/EES. 2005d. SFPUC Lead Source Replacement Case Study. Draft Report for AwwaRF Project No. 3018. HDR Engineering, Bellevue, WA.

Hertz-Picciotto, I., M. Schramm, M. Watt-Morse, K. Chantala, J. Anderson, and J. Osterloh. 2000. "Patterns and determinants of blood lead during pregnancy." *Am J Epidemiol.* 152:829-837.

Heumann, D. 2005. "LADWP Non-lead Brass Program." Email to T. Case. 9.30.05.

Ingleson, H., A.M. Sage and R. Wilkinson, 1949. "The Effect of the Chlorination of Drinking Water on Brass Fittings." *J. Inst. Water Engineers* (Br.), 3:81.

International Agency for Research on Cancer. 2006. Inorganic and Organic Lead Compounds. IARC Monogr. Eval. Carcinogen. Risks Hum., 87 (In preparation). http://monographs.iarc.fr/ENG/Meetings/vol87.php

Kim, J.G, S.B. Jung, and O.H. Kwon. 2001. "Technical Note: Dealloying Behavior of Unleaded Brasses Containing Bismuth in Potable Water." *Corrosion*, V. 57, N. 4 291-294.

Kimbrough, D. 2001. "Brass Corrosion and the LCR Monitoring Program." *Journal American Water Works Association*. Vol. 93, pp. 81-91, 2001.

Kuyucak S. and M. Sahoo. 1996. "A Review of the Machinability of Copper-Base Alloys." *Canadian Metallurgical Quarterly*, Vol.35, No. 1, pp. 1-15, 1996.

Lanphear, B. P., K. Dietrich, P. Auinger, and C. Cox. 2000. "Cognitive deficits associated with blood lead concentrations <10 microg/dL in US children and adolescents." *Public Health Rep* 115, 521-9.

Laprade D. 2005. Massachusetts Department of Environmental Protection, Western Regional Office. Personal Communication, 2005.

Larson, T.E., R.M. King and L. Henley, 1956. "Corrosion of Brass by Chloramine." *Jour. AWWA*, 84-88 (January 1956).

Lucey, V.F. "The Mechanism of Dezincification and the Effect of Arsenic, 1965." *Br. Corrosion Journal*. V. 1. N. 1 53-59.

Lytle, D.A. and M.R. Schock. 1996. S*tagnation Time, Composition, pH and Orthophosphate Effects on Metal Leaching from Brass (EPA/600/R-96-103)*. National Risk Management Research Laboratory, Office of Research and Development, USEPA, Cincinnati, Ohio September 1996.

Maas, R. 2002. *Measurements of Lead, Bismuth and Selenium Discharge from Water Service Valves and Fittings Manufactured From SeBiLOY® (Envirobrass®) Brass Alloy*. UNC-Asheville Environmental Quality Institute, Technical Report #02-105. December 2002 Accessed at http://orgs.unca.edu/eqi/LFPub2.doc on 15/10/2005.

Maas, R.P., S.C. Patch, and M. Lagasse. 2002. *Measurements of Lead, Bismuth and Selenium Discharge From Water Service Valves and Fittings Manufactured From SeBiLOY® (Envirobrass®) Brass Alloy.* UNC-Asheville Environmental Quality Institute, Technical Report #02-105. Asheville, NC.

Maas, R. and S. Patch. 2004. "Update on Research Findings and Regulatory/ Legal Activities Regarding Tap Water Lead Exposure From Traditional Leaded-Brass and "Non-lead" Type Plumbing Parts." *Journal of the New England Water Works Association.* In press for March 2004 Issue. Accessed at http://orgs.unca.edu/eqi/LFPub3.htm on 25/10/2005.

Mass, R. 2003. *Assessment of the Ford Meter Box Company Wash System for Reducing Lead Discharge from Leaded Brass Parts.* UNC-Asheville Environmental Quality Institute, October 2003.

McCLellan, C. 2005. Email to K. Martel. 10.14.05.

McClellan, C. 2006. Email to A. Sandvig. 6.2.06.

Menke, A., P. Muntner, V. Batuman, E. Silbergeld, and E. Guallar. 2006. "Blood Lead Below 0.48 μmol/L (10 μg/dL) and Mortality Among US Adults." *Circulation* 114:1388-1394.

Moriarty, K. 2005. Personal communication with K. Martel (HDR/EES). 9/27/05.

Nawrot, T. and J. Staessen. 2006. "Low-Level Environmental Exposure to Lead Unmasked as Silent Killer." [Editorial] *Circulation* 114:1347-49.

NDT. 2005. Nondestructive Testing (NDT) Resource Center, Standards and Specifications. Accessed at http://www.ndt-d.org/GeneralResources/Standards/standards.htm on 11/10/2005.

Needleman, H.L. and C.A. Gatsonis. 1990. "Low-Level Lead Exposure and the IQ of Children: A Meta-Analysis of Modern Studies." *J Am Med Assn*, 263(5):673-678.

NSF. 2003. *NSF/ANSI 61-2003e. Drinking Water System Components – Health Effects.* NSF International. Ann Arbor, Michigan.

OECD. 1994. *OECD Workshop on Lead Products and Uses, Session F: Faucets/Water Coolers, Chairman's Report, Canada 1994.* Accessed at www.oecd.org/dataoecd/36/28/1937028.pdf on 10/01/2005.

OEHHA. 2005. "Proposition 65 Information". Office of Environmental Health Hazard Assessment (OEHHA), California. Accessed at http://www.oehha.ca.gov/PROP65.HTML on 11/02/2005.

Paige J. and B. Covino. 1992. "Leachability of Lead from Selected Copper-base Alloys". *Corrosion.* Vol. 48, No 12, pp. 1040-1046, 1992.

Peters, D. T. 1995. "Bismuth-Modified Cast Red Brasses to Meet U.S. Drinking Water Standards." *Copper 95; Third International Conference, Santiago Chile, November 26-29, 1995 Plenary Lectures, Economics, Applications, and Fabrication of Copper. Vol. 1*, pp. 301-322. Canadian Institute of Mining, Metallurgy, and Petroleum. Montreal, Canada.

Pierson, G.L. 2006. Personal Communication with HDR project manager for water quality study with Seattle Public Schools (October 21, 2005 and April 5, 2006).

Pugh, E.N., W.G. Montague, and A.R.C. Westwood, 1966. "Stress Corrosion Cracking of Copper." *Corrosion Science*, V. 6 345-346.

Rao, T.S. and K.V.K. Nair, 1998. "Microbiologically influenced stress corrosion cracking Failure of admiralty brass condenser tubes in a nuclear power plant cooled by freshwater." *Corrosion Science.* V. 40, N. 11 1821-1836.

Sadayappan M., L. Whiting and M. Sahoo. 2003. "Lead-Free Envirobrass® - An Environmentally Friendly Ecomaterial". *42nd Annual Conference of Metallurgists of CIM*. Vancouver, British Columbia, Canada, August 2003.

Sadayappan K. 2005a. Materials Technology Laboratory of CANMET Canada. Personal Communication 2005.

Sadayappan K. 2005b. "Mitigating Lead in Copper Alloys for Drinking Water Applications". *Foundry Management and Technology*, pp. 59-62, March 2005.

Sahoo, M., J. Dion , S. Kuyucak and M. Shehata. 1993. *A Feasibility Study of the Production of Free-Machining Brass without Lead*. Metals Technology Laboratories, CANMET Canada, March 1993.

Sahoo, M, L. Whiting and M. Sadayappan. 1999. "A Review of Bismuth and Selenium Modified Copper Alloys for Plumbing Applications." *Presented at Copper 99-fourth International Conference, Phoenix, October 10-13, 1999.*

Sambo Copper Alloy Co., 2005. ECO BRASS® Information. Accessed 11/02//2005 at http://www.sambo.co.jp/products_e/ecobrass.htm

Schwartz, J. 1994. "Low-Level Lead Exposure and Children's IQ: A Meta-Analysis and Search for a Threshold." *Environmental Research*, 65:42-55.

Seattle Public Schools. 2006. Water Quality at Seattle Public Schools. Accessed Oct 31, 2005 and April 5, 2006. www.seattleschools.org/area/facilities/WaterQuality/water_quality_update.htm

Selover, C. 2005. "Manufacturers Concerns –Historical and Future Perspective". Presented at the *U.S. EPA Lead and Copper Rule Workshop: Lead Content in Plumbing Fittings and Fixtures*. July 26-27, 2005. U.S. Environmental Protection Agency, Washington, D.C.

Showman, R. 1994. "Experiences with lead-free brass in the manufacture of Cast Valves and Fittings." *Proceedings American Water Works Association Annual Conference, New York, June 19-23, 1994*. AWWA, Denver, CO.

Sundberg, R, R. Holm, S. Hertzman, B. Hutchinson and E. Lindh-Ulmgren, 2003. "Dezincification (DA) and intergranular corrosion (IGA) of brass-influence of composition and heat treatment." *Metal*. V. 57 721-731.

Sundberg, R, 2006. Personal Communication, February 9th.

Thompson, D.H., and A.W. Tracy, 1949. "Influence of Composition on the Stress-Corrosion Cracking of Some Copper-base Alloys." *Metals Transaction*. February 100-109.

Twarog, D. 1995. "AFS Research: Modified Red Brass With Bismuth and Selenium Research Results." *Proceedings AFS 99th Casting Congress*, Missouri, April 1995.

U.S. Centers for Disease Control and Prevention. 2005. Preventing Lead Poisoning in Young Children. Atlanta, GA http://www.cdc.gov/nceh/lead/ACCLPP/meetingMinutes/lessThan10MtgMAR04.pdf

U.S. Environmental Protection Agency. 1987. "Amendments to the Safe Drinking Water Act." *52 FR 20674*, June 2, 1987.

U.S. Environmental Protection Agency. "1988 Lead Contamination Control Act." *Public Law 100-572*. October 31, 1988.

U.S. Environmental Protection Agency. "1996 Amendments to the Safe Drinking Water Act". *Public Law 104-182* 104th Congress.

U.S. Environmental Protection Agency. 1997. "Interpretation of New Drinking Water Requirements Relating to Lead Free Plumbing Fittings and Fixtures." *Fed. Reg.* Vol. 62, No. 163, 44684, Aug. 22, 1997.

U.S. Environmental Protection Agency. 2004a. *USEPA Lead and Copper Rule Workshop 2: Monitoring Protocols Summary. May 12-13, 2004.* http://www.epa.gov/safewater/lcrmr/pdfs/summary_lcmr_review_monitoring_workshop_summary_05-04.pdf . Accessed 9/16/05.

U.S. Environmental Protection Agency. 2004b. USEPA Lead and Copper Rule Workshop 1: Simultaneous Compliance Summary, May 11-12, 2004. Accessed at http://www.epa.gov/safewater/lcrmr/pdfs/summary_lcmr_review_simultaneous_compliance_workshop_05-04.pdf on 11/02/2005.

U.S. Environmental Protection Agency. 2005a. National Review of LCR implementation and Drinking Water Lead Reduction Plan. Accessed 9-19-2005. http://www.epa.gov/OGWDW/lcrmr/lead_review.html

U.S. Environmental Protection Agency. 2005b. USEPA Lead and Copper Rule Workshop: Lead in Plumbing, Fittings, and Fixtures. July 26-27, 2005. Accessed 9/19/2005 www.epa.gov/OGWDW/lcrmr/pdfs/summary_lcmr_review_school_plumbing.pdf

U.S. Environmental Protection Agency. 2005c. Technology Transfer Network Air Toxics Website, Selenium Compounds. Accessed 11-16-05 http://www.epa.gov//ttn/atw/hlthef/selenium.html

U.S. Environmental Protection Agency. 2005d. Office of Ground Water and Drinking Water Website. Consumer Fact Sheet – Copper. http://www.epa.gov/safewater/dwh/c-ioc/copper.html. Accessed 11-09-2005.

U.S. Environmental Protection Agency. 2005e. Office of Ground Water and Drinking Water. Consumer Fact Sheet – Nickel. http://www.epa.gov/safewater/dwh/c-ioc/nickel.html. Accessed 11-9-2005.

U.S. Environmental Protection Agency. 2005f. Office of Ground Water and Drinking Water. Consumer Fact Sheet – Lead. http://www.epa.gov/safewater/dwh/c-ioc/lead.html. Accessed 11-9-2005.

U.S. Geological Survey. 2005a. *Mineral Commodity Summaries*. U.S. Geological Survey. U.S. Department of Interior. January 2005.

U.S. Geological Survey. 2005b. *Mineral Industry Surveys.* U.S. Geological Survey. U.S. Department of Interior. September 2005.

Viola, D.W. 2002. "How Important are Model Codes and Product Standards?" *PMI News*, July/August issue. www.pmihome.org/techedu-articles-2002-jul.shtml .

Viola, D.W. 2005. Email to Anne Sandvig. December 21, 2005.

Weed, R. 2005. "Replacing Leaded Brass Alloys: Issues, Concerns, Opportunities. Presented at Plumbing Manufacturers Institute Fall Meeting, Washington, D.C.

Weil, E. 2005. "California's experience using Proposition 65 to Reduce Lead in Plumbing Fixtures". Presented at USEPA Lead and Copper Rule Workshop: Lead in Plumbing Fittings and Fixtures, July 26-27, 2005. Accessed at http://www.epa.gov/safewater/lcrmr/pdfs/summary_lcmr_review_school_plumbing.pdf on 10/02/2005.

World Health Organization (WHO). 1993. *Guidelines for Drinking Water Quality, 2nd Edition, Volume 1*, pp. 49-50. Geneva, Switzerland: WHO. www.who.int/water_sanitation_health/diseases/lead/en

World Health Organization (WHO). 2005. International Program on Chemical Safety (IPCS) Monographs and Toxicological Data. http://www.who.int/ipcs/publications/training_poisons/basic_analytical_tox/en/index8.html. Accessed 11-9-2005

ACRONYMS AND ABBREVIATIONS

A.L.	Action Level
ASME	American 'Society of Mechanical Engineers
AFS	American Foundrymen's Society
ASTM	American Society for Testing Materials
AG	Attorney General
ASCE	American Society of Civil Engineers
ASDWA	Association of State Drinking Water Administrators
AWWA	American Water Works Association
AwwaRF	Awwa Research Foundation
BBIM	Brass and Bronze Ingot Manufacturers
BDL	Below Detection Level
CA	California
CDA	Copper Development Association
DCWASA	District of Columbia Water and Sewer Authority
dL	Deciliter
$/lb	Dollars per pound
EPA	Environmental Protection Agency
GCWW	Greater Cincinnati Water Works
IAPMO	International Association of Plumbing and Manufacturing Officials
IARC	International Agency for Research on Cancer
ICP-MS	Inductively Coupled Plasma Mass Spectrometry
ISRI	Institute of Scrap Recycling Industries
LCR	Lead and Copper Rule
LCCA	Lead Contamination Control Act
LADWP	Los Angeles Department of Water and Power
MCL	Maximum Contaminant Level
ug/dL	micrograph per decilaiter
mg/L	Milligram per liter
MPa	Milli-Pascal
NSPC	National Standard Plumbing Code
NDT	Non-destructive testing
NFFS	Non-ferrous Founders' Society
NSF/ANSI	NSF International/American National Standards Institute
OEHH	Office of Environmental Health Hazard Assessment. California, U.S.
O&M	Operations and Maintenance
OECD	Organisation for Economic Co-operation and Development.
PMI	Plumbing Manufacturers Institute
PPB	Parts per Billion
SDWA	Safe Drinking Water Act
SFPUC	San Francisco Public Utility Commission
SPS	Seattle Public Schools
UPC	Uniform Plumbing Code

U.S.	United States
USCDC	United States Centers for Disease Control
USEPA	United States Environmental Protection Agency
USGS	United States Geological Society
VPI	Virginia Polytechnic Institute and State University
WHO	World Health Organization
Zn	Zinc

IWA Publishing's authorised EU representative for General Product Safety Regulations is Diane D'Arras, 15 rue Duret, 75116 Paris, France, e-mail: safety@iwap.co.uk.

Printed and bound by CPI Group (UK) Ltd, Croydon, CR0 4YY
26/09/2026
02223459-0001